THE MAN IN THE MONKEYNUT COAT

T0202211

THE MAN IN THE MONKEYNUT COAT

William Astbury and How Wool Wove a Forgotten Road to the Double-Helix

Kersten T. Hall

School of Philosophy, Religion and History of Science, University of Leeds

OXFORD
UNIVERSITY PRESS

Great Clarendon Street, Oxford, OX2 6DP,
United Kingdom

Oxford University Press is a department of the University of Oxford.
It furthers the University's objective of excellence in research, scholarship,
and education by publishing worldwide. Oxford is a registered trade mark of
Oxford University Press in the UK and in certain other countries

© Kersten T. Hall 2014, First published 2014, First published in paperback 2022

The moral rights of the author have been asserted
Impression: 1

Published in the United States of America by Oxford University Press
198 Madison Avenue, New York, NY 10016, United States of America

British Library Cataloguing in Publication Data
Data available

Library of Congress Control Number: 2022934227

ISBN 978–0–19–876696–4

DOI: 10.1093/oso/9780198766964.001.0001

Printed and bound by
CPI Group (UK) Ltd, Croydon, CR0 4YY

To Edward, Matthew, my wife Michelle, my father Gordon and
the memory of my mother Gisela

Acknowledgements

This book could not have been written without the help of a large number of people, and if I have omitted any names I sincerely apologise. I thank Professor Greg Radick, School of Philosophy, Religion and the History of Science, University of Leeds, not only for first suggesting to me that Astbury might be an interesting subject for study but also for persuading me that counterfactual history is more than just a handy plot device for sci-fi stories. In addition, I extend my thanks to everyone else who has given their interest, support, and help to this work, in particular Professor Tony North, Dr Chris Hammond, and Professor Denis Greig, University of Leeds for valuable discussions about Astbury and the Braggs; Dr John Lydon for showing me a fibre of 'Ardil'; Dr Keith Parker for taking the time to share his recollections of life in Astbury's laboratory at 9 Beechgrove Terrace; Jim Garretts for help in obtaining this recording and Astbury's 1942 lecture 'Science Lifts the Veil'; Jenifer Glynn for kindly taking the time to read a draft of the manuscript and not only offering helpful feedback but also providing photographs of her sister, Rosalind Franklin; all the team in Special Collections at the Brotherton Library, University of Leeds, particularly Chris Sheppard, Karen Mee, Fiona Gell, Jonathon Horn, and Matt Dunne; Katy Thornton, Head of Special Collections for kindly granting permission to cite the sources and photographs in the Astbury papers at the University of Leeds; Nick Brewster of the Central Records Office, University of Leeds; Dr Peter McHugh, Dr Cordelia Sealy, and Professor Jordan Goodman for feedback on an early draft and valuable advice on approaching publishers; all staff and students in History and Philosophy of Science at Leeds, particularly Professor Graeme Gooday, Dr Jon Topham, Dr Adrian Wilson, and Dr Annie Jamieson for information about the adoption of X-ray technology; Elwyn Beighton's family, particularly Mrs Suzannah Sanderson and her daughter Gemma, for their support and interest; Professor Elspeth Garman, University of Oxford; Dr Ruth Seymour; Professor Robert Olby for some very insightful discussions about Astbury; Michael P. Miller, Manuscripts Processor at the American Philosophical Society for kindly granting access to Erwin Chargaff's papers; Chris Petersen, Oregon State

University, Special Collections and Archives for granting access to the Ava and Linus Pauling Papers; Madelin Terrazas, Archives Assistant, Churchill Archives Centre, Churchill College for help in locating Astbury's correspondence with John Turton Randall; Emily Dezurick-Badran, Manuscripts Archive, Cambridge University Library; Hannah Westall, Archives, Girton College, Cambridge for help in tracing information about Florence Bell; Crestina Forcina and Natalie Walters, Wellcome Library for permission to cite letters by Francis Crick; Jonathon Smith, University of Cambridge Library for help in searching the papers of Richard Synge; the National Library of Australia for permission to cite newspaper reports of William Bragg's X-ray demonstrations in Adelaide; Dr Tony Slavotinek, South Australian Medical Heritage Association for permission to cite correspondence relating to Samuel Barbour; Lianne Smith, Archives, King's College, London for granting permission to reproduce 'Photo 51'; Jill Winder, ULITA Archive, University of Leeds; Cold Spring Harbor Laboratory Press for permission to use the 'pile of pennies' image; the Rockefeller Institute for permission to use the photograph of Oswald Avery; Erica Peacock, National Archives, Kew; Stella Butler and Beccy Shipman, University of Leeds Library for permission to reproduce photographs of bacterial flagella taken by Elwyn Beighton and photographs of sodium thymonucleate taken by Florence Bell in her PhD thesis; Christine Holdstock for useful information on the Wool Industries Research Association; Louise North, BBC Written Archives Centre for help with transcripts of Astbury's radio broadcasts; Dr Lee Hiltzik, Rockefeller Archives Center; the Astbury Centre for Structural Molecular Biology, University of Leeds for support and interest, particularly Professor S. Radford, Professor A. Berry, Dr Arwen Pearson, Dr Bruce Turnbull, and Professor Adam Nelson for interesting discussions about Astbury's legacy today; Professor Andre Authier for helping me to approach Oxford University Press and Doctor James McHugh, University of Southern California for his invaluable advice on preparing the synopsis of a manuscript; Ross at Leeds Central Library for help in locating the original *Yorkshire Evening Post* cartoon showing the monkeynut coat; Julie Mahoney at Huddersfield Library; Dr Isabel Whitehouse, University of Leeds for useful information on protein misfolding diseases; Imogen Clarke for her dissertation on Astbury's University of Nottingham, for the loan of *The Man in the White Suit*, and Dr Charlotte Sleigh of the University of Kent, Dr Peter Morris of the Science Museum London, Dr Amy Sargeant, and Mr James Pepper for insightful discussion of the film; Ian Rothwell and the Friends of the Thackray Museum, Leeds City Museum, and Dr Emily Winterburn and Dr Claire Jones of the Museum of the History of Science, Technology and Medicine, University of Leeds for inviting me to give the public lectures from which the idea for this book came; the friends and family who supported these talks, particularly Professor A. Whitehouse, University of Leeds for kindly providing the reagents and equipment for the practical demonstrations that accompanied some of these lectures, and Joe, Pariz, and Emily, the undergraduate students who helped with them. Also thanks to Professor G. E. Blair, University of Leeds for first drawing my attention to the work of Archer Martin and Richard Synge. I also thank the many other friends who have listened politely over the years while I enthused to them about the importance of Astbury and his work. I am also grateful to Dr David A. Harris, St Anne's College Oxford, who first prompted me nearly 26 years ago to think about how a polypeptide chain might be formed; Ashley Stokes and Robin Jones for invaluable feedback at the early stage; and my editors Sonke Adlung and Jessica White at OUP for their interest, enthusiasm, and support for this work. And special thanks to Astbury's grandson William and all his family for their support, interest, and help in providing access to family documents.

Since the original publication of this book in 2014, some of those who helped me write it are sadly no longer with us. So I'd just like to pay tribute to the memory of Astbury's former PhD student Professor Laszlo Lorand for kindly sharing his recollections of his time in Astbury's lab, his thoughts on Elwyn Beighton's X-ray images of DNA, and his permission to use his audio recording of Astbury's 1953 lecture 'How to Swim With a Molecule For a Tail.' Also to Professor Monty Lososwky for sharing his wonderful memory of having been the junior doctor who Astbury ordered to go to the medical library and read up on the molecular mechanism of blood clotting before treating him, and Dr. Vivian Wyatt for his useful discussions on the history of molecular biology.

Lastly, of course, a huge thanks to my wife Michelle and sons Edward and Matthew for their unwavering patience and feigning polite interest as I have talked incessantly at the dinner table about wool, DNA, and why a coat made from monkeynuts matters over the past few years.

Contents

Introduction to the second edition

It's a pity that William Astbury was not as good a comedian as he was a physicist. Had his jokes been funnier, he might well have found himself sharing a Nobel Prize for one of the biggest discoveries of the twentieth century. Instead, his name is today unknown except to a small group of academic historians.

But in 1951, when Astbury attended a scientific meeting in Naples, things were very different. Astbury was a scientific giant and one of the leading lights in the emerging new science of 'molecular biology', which sought to explain the complexity of living systems in terms of molecular shape. Yet this fame came from humble and unlikely beginnings. Using the scattering of X-rays to study the molecular structure of wool fibres for the local textile industries of West Yorkshire, Astbury had gained crucial insights into the nature of proteins—Nature's nanomachines—that carry out a vast array of functions within living cells. The studies that began with work on the humble wool fibre eventually led to the very first structural studies of another white, stringy fibre—DNA, the material of heredity.

Thirteen years before Astbury attended the Naples meeting, his research assistant Florence Bell had made the very first X-ray studies of DNA. In so doing, she showed for the very first time that X-rays could be used to reveal the regular, ordered structure of DNA. As a result of this work, one particular delegate at the Naples conference was very keen to speak to Astbury.

Maurice Wilkins was a physicist who had worked on the Manhattan Project, but after being horrified by the devastation the bombs unleashed upon Hiroshima and Nagasaki, Wilkins now turned his attention to biology—and to solving the structure of DNA. But another young scientist at the meeting also had DNA on his mind: young American James Watson was convinced that the secrets of the gene lay in solving the structure of DNA and that the X-ray methods used by Wilkins and Astbury offered the means to do this.

But which of the two men should he approach—Astbury, the elder statesman, or Wilkins, the newcomer to the field? Wilkins himself was delighted—coming to Naples provided him with the chance to finally meet Astbury: 'I was very glad to meet him . . . He liked making little jokes . . .'.[1] But while Astbury's sense of humour may have endeared him to Wilkins, it had a very different effect on Watson. Wilkins liked Astbury 'because he seemed a real human being', but for Watson that was exactly the problem. Instead of meeting an intellectual X-ray titan as he had hoped, when Watson saw Astbury for the first time, the over-riding impression was of a 'has-been,' whose best years were long behind him. He seemed to have little now to offer other than a few 'off-colour jokes' told while drinking copious amounts of whisky.[2]

Having reached this conclusion, Watson moved on and introduced himself instead to Wilkins. The rest, as they say, is history. Two years after the meeting in Naples, Watson and his colleague Francis Crick published their double-helical structure of DNA, for which they and Maurice Wilkins were jointly awarded the 1962 Nobel Prize in Physiology or Medicine.

We can speculate about how differently history might have unfolded had Watson gritted his teeth, feigned laughter at Astbury's flat jokes, and shown some interest in his work on DNA. While Astbury was in Naples doing a bad impression of a stand-up comedian, back in Leeds his research assistant Elwyn Beighton was about to obtain a stunning new X-ray image of DNA that contained a vital clue to unravelling its structure. A year later, an identical image taken by the crystallographer Rosalind Franklin and her PhD student Raymond Gosling would have a dramatic effect when shown to Watson for the first time. Instantly recognising that the pattern formed by the scattered X-rays could only have been made by a helical molecule, Watson said 'my mouth fell open and my pulse began to race.'[3] Yet had he been able to overlook Astbury's bad jokes in Naples, he might well have been sufficiently intrigued to pay a visit to Leeds, where Beighton's photo would have sent his pulse racing a year earlier and set him and Astbury on the path to the Nobel prize.

But why should we care about events that didn't happen? And why read - or even for that matter, write - a book about a forgotten scientist who didn't solve the structure of DNA?

Firstly, sometimes, by exploring what *might* have happened we can actually gain some important new insights into what actually *did* happen. And this new understanding can be invaluable in shaking us out of a complacent sense that somehow events *had* to unfold in the way that they did. But secondly, and more importantly, Astbury did more than just blaze the trail for Watson, Crick, and Wilkins with their work on DNA.

Astbury's greatest scientific legacy took the form of a rather unusual overcoat he wore—woven not from wool, cotton, or other conventional textile fibres, but rather from edible proteins found in ordinary monkeynuts. This textile was produced using an ingenious act of molecular origami and, while the coat itself has long since rotted away in some dusty cupboard, the idea behind its creation has endured, and this was Astbury's lasting vision—that molecules and their shapes are the key to unravelling the mysteries of living systems. It's an idea that allowed Astbury to explain how muscles contract and bacteria can swim and, in the time since this book was first published, it has taken on a powerful new relevance to all those of us who have benefited from a jab against Covid-19. This final example alone surely makes the story of the monkeynut coat, and the man who wore it, worth telling.

1

A Picture Speaks a Thousand Words

Maybe he was just an avid crossword enthusiast. That might explain why the young man in the corner of the railway carriage was scribbling with a quiet and focused intensity on his newspaper. A fellow passenger glancing across at him might have quickly concluded that he was either engaged in a concentrated effort to complete the day's clues before reaching his final destination or perhaps distracting his mind from the discomfort of sitting in an unheated railway carriage, late at night in the middle of a particularly miserable English winter. Either way, they would have been at least partially right. For he was indeed working out the solution to a puzzle. But it was one that was far bigger than any crossword, and one for which the scribbles on his newspaper would eventually lead to a Nobel Prize.

It was late January in 1953. The coming year would see a new young queen crowned in Britain, Everest conquered, and Joseph Stalin dead. But of all these events, perhaps the most momentous was a short paper that appeared in the scientific journal *Nature* of April that year. Just over a page long, it described a proposed structure for deoxyribonucleic acid (DNA), the molecule that is central to the inheritance of biological information from one generation to the next.[1] It was a discovery that not only transformed our understanding of biology, but also brought with it the power to alter living systems at the molecular level. It also earned its authors, the Cambridge scientists James Watson and Francis Crick, the Nobel Prize—the most coveted award in science.

Of all the clues to solving this mystery, one of most important lay in the scribbles that James Watson was making in the margin of his newspaper as he travelled home that night on the train from London to Cambridge. For, rather than filling in the crossword, he was instead sketching a pattern of spots that formed the distinctive shape of a large letter 'X'. What was this pattern? Where had it come from, what did it mean, and why was Watson so keen to scribble it down as the train rattled back across the fens to Cambridge?

Two years earlier, Watson had experienced an epiphany. He had been at a scientific conference in Naples where the physicist Maurice Wilkins had shown a series of photographs that intrigued him. Working at King's College, London, Wilkins and his PhD student Raymond Gosling had taken a number of photographs that captured the patterns made when X-ray beams were scattered by molecules of DNA. The ordered patterns of spots on Wilkins' pictures were strong evidence that DNA contained at least some regions that were crystalline—in other words, it was made of a regular repeating structure. The regularity suggested by Wilkins' X-ray pictures left Watson convinced that not only could the structure of DNA be solved, but also that within this structure lay the secret of how the molecule passed on genetic information. There and then, Watson made up his mind that solving the structure of DNA was the task to which he would devote himself.[2]

Inspired by the X-ray pictures of DNA that Wilkins had shown in Naples, Watson moved to the Cavendish lab in Cambridge where he and his colleague, Francis Crick began to construct a theoretical model of what the molecule might look like. But progress did not go smoothly. Having invited Wilkins and his colleague from King's, the crystallographer Rosalind Franklin, to come to Cambridge to view their first model, they were somewhat humiliated when Franklin quickly pointed out a number of basic schoolboy errors in their chemistry. As a result, the head of their department, Lawrence Bragg, imposed a moratorium on their work on DNA and insisted that they return to their original respective areas of research—for Crick, the structure of the blood protein haemoglobin, and for Watson, the structure of viruses. To add to their frustration at being banned from further work on DNA, news reached them in late 1952 that the eminent American chemist Linus Pauling was about to publish a paper on the structure of DNA. Pauling was the world's greatest structural chemist, who, only a year earlier, had beaten Bragg's team to solving one of the basic puzzles in biochemistry—how protein molecules are coiled up into helices. It was a defeat from which Bragg was still smarting and the possibility that Pauling was about to scoop the structure of DNA as well was a particularly painful prospect for the Cambridge team.

Yet when Watson obtained a copy of Pauling's manuscript, he breathed a sigh of relief. From a quick glance at the data, what he knew of chemistry told him that Pauling's proposed model for the structure of DNA could not possibly be correct. There might still be time for him and Crick to claim the DNA structure for Cambridge. Spurred on by this, he got on the train to London to find out what Wilkins and Franklin at King's made of Pauling's paper.

By the time that Watson made his trip to London clutching Pauling's manuscript, however, Wilkins and Franklin were barely on speaking terms. Wilkins may have initiated X-ray studies on DNA at King's, but they were now largely the domain of Franklin. Still only in her early thirties, Franklin had quickly established herself as a leading expert in the field of X-ray crystallography—a technique that used the scattering of X-rays to deduce the structure of molecule (Figure 1). Depending on their particular shape and repeating arrangement in space, molecules scattered, or 'diffracted', the X-rays in different ways that could be recorded as a pattern of black spots on an X-ray film. By careful analysis and the application of particular mathematical methods, the physical shape of the molecule could then be deduced from

Figure 1 Rosalind Franklin enjoying the scenery of Tuscany.
Photograph by Vittorio Luzatti. Reproduced with permission of Dr Jenifer Glynn.

these patterns of spots on the film. Having worked for several years in Paris using this method to study the structure of carbon-based compounds such as coal, Franklin had now come to King's where her new supervisor, the physicist John T. Randall, had won support from the Medical Research Council (MRC) to set up a new research unit for what he called 'Biophysics'.

Thanks to poor communication and leadership by Randall, Franklin had arrived at King's in January 1951 with the mistaken belief that X-ray work on DNA would now be her sole preserve and that Wilkins would be returning to his earlier research using microscopes. As a consequence of this misunderstanding, Franklin and Wilkins' working relationship was destroyed and, rather than work together in what could and should have been a very successful collaboration, each of them now continued their research in isolation from the other.

As a result of this awkward situation, when Watson arrived at King's, he sought out Franklin and Wilkins individually to ask their opinion on Pauling's manuscript and found, as he had hoped, the confirmation that Pauling was wrong. But he also found far more. Out of Franklin's earshot, Wilkins mentioned to Watson that she and her PhD student Raymond Gosling had obtained an interesting new X-ray photograph of DNA that they had labelled simply as 'Photo 51'. As Wilkins had been told about the picture by Gosling, who had been his own PhD student before Franklin had taken over this role, he saw no problem in showing Watson the image. Moreover, Wilkins knew that Franklin would soon be giving up her work on DNA as she was due to leave King's to take up a new post at Birkbeck College. In his autobiography, however, Wilkins said that, with hindsight, showing the picture to Watson was a decision that he had come to regret.[3] And well he might.

At first sight, the picture would probably not look out of place hanging in a gallery of contemporary art. It shows a number of smeared, blurry, black spots almost reminiscent of ripples on the surface of a pond, but what is immediately striking is the way in which these spots form the distinct shape of a cross (Figure 2). Watson's reaction on seeing this cross pattern for the first time was dramatic:

> The instant I saw the picture my mouth fell open and my pulse began to race . . . the black cross of reflections which dominated the picture could only arise from a helical structure.[4]

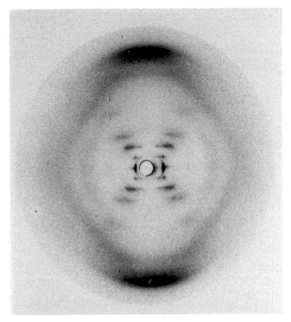

Figure 2 X-ray diffraction photograph of 'B' form DNA taken in 1952 by Rosalind Franklin and her PhD student Raymond Gosling showing the striking black cross that was immediately recognised by James Watson as being characteristic of a helical molecule. Known as 'Photo 51', the image was published in Franklin, R., and Gosling, R. (1953). Molecular configuration in sodium thymonucleate. *Nature*, 171, pp. 740–741.

Reproduced with permission of King's College Archives.

What set Watson's pulse racing was his knowledge that this kind of cross pattern could only be made by X-rays scattered from a molecule that was coiled up into a helical shape. Sitting on the train that evening, he scribbled the details of the photograph down on his newspaper ready to show Crick once he arrived back in Cambridge. Franklin and Gosling's photograph had given Watson not only the vital experimental data that he and Crick needed to confirm their theoretical models, but also a powerful piece of evidence to persuade Lawrence Bragg to allow them to resume their research into DNA. Not wishing to be beaten yet again by Pauling, Bragg duly obliged. In a matter of weeks, Watson and Crick had completed building a new model of DNA and submitted their now famous paper to *Nature*, for which, nine years later, they would receive the Nobel Prize. Understanding the structure of DNA gave insights into some of the deepest questions in biology. The double-helical twist of the molecule had far more than just aesthetic appeal, for it explained one of the fundamental properties of the genetic molecule—its ability to replicate. In Watson and Crick's model, DNA was composed of two chains that twisted around each other, made of repeating units of the sugar deoxyribose linked to phosphate (Figure 3). Jutting out perpendicular to each sugar phosphate chain were chemicals called bases that came in four varieties—adenine, guanine, cytosine, and thymine. The crucial feature of the double-helical shape was that it enabled the bases on opposite chains to pair up with each other in a very precise way.

Figure 3 Watson and Crick's double-helical model of DNA. The significance of the double-helical structure is that it allows the bases (A, C, G, T) on opposite strands of the molecule to pair up in a very specific way—A with T, and G with C. During replication, the two strands separate and, thanks to the specific rules of base-pairing, each strand can act as a template for the synthesis of a new, complementary strand.

Reprinted by permission from Macmillan Publishers Ltd: *Nature*, Watson, J. D. and Crick, F. H. C. (1953). Molecular structure of nucleic acids: a structure for deoxyribosenucleic acid. *Nature*, 171, pp. 737–738.

This was significant because this specific pairing between bases on opposite chains not only held the molecule together, but also explained how the molecule could make copies of itself, allowing biological traits to be passed from one generation to the next.[5] But Watson and Crick's model also did much more than answer theoretical questions. In the early 1970s, it became evident that understanding the molecular structure of DNA might allow it to be deliberately altered and manipulated. Nucleic acids from one organism could be physically combined with those from another, entirely unrelated, organism to give rise to hybrid molecules. It was this new technology of recombinant DNA that lay at the heart of the biotechnology industry that grew up in the mid-1970s, of which the production of human insulin by bacteria was a spectacularly successful example. Nor was the impact of understanding the structure of nucleic acids confined only to the biomedical world—it also lay at the heart of the DNA fingerprinting techniques used by the police to solve crimes.

Little wonder, then, that a plaque on the wall outside King's College on the Strand in London where Franklin worked hails 'Photo 51' as being 'one of the most important photographs in the world'. Sadly, however, Rosalind Franklin never lived to see the full implications of her work, for, aged only 37, she died in 1958 of ovarian cancer. Her sister Jenifer Glynn has since been reported as saying that, had Franklin lived, she would have 'exploded with fury' to discover that Watson and Crick had used her data without her knowledge.[6] For when Watson and Crick's discovery was first published in *Nature*, in April 1953, she received only the following, rather modest, acknowledgement:

> We have also been stimulated by a knowledge of the general nature of the unpublished results and ideas of Dr. M. H. F. Wilkins, Dr. R. E. Franklin and their co-workers at King's College, London.[7]

To say that they had been 'stimulated' by Franklin's work is rather an understatement. As Crick himself conceded in a letter written to the French biologist Jacques Monod in December 1961, 'the data which really helped us to obtain the structure was mainly obtained by Rosalind Franklin'.[8] It is important to note that the information in 'Photo 51' alone was not enough to allow Watson and Crick to solve the structure of DNA. They also needed X-ray data contained in a report that Franklin had written to the MRC and from which Watson and Crick were able to deduce that the molecule consisted of two chains running in opposite directions. But 'Photo 51' and the clue it contained nevertheless had a powerful impact upon them. In an address at the inauguration of the Harvard Center for Genomic Research in 2001, Watson publicly acknowledged the role that Franklin's picture had played in his success, saying: 'it was that the Franklin photograph was the key event. It was, psychologically, it mobilised us . . .'.[9]

Yet, despite this, when Crick and Watson were awarded the Nobel Prize in 1962, neither of them gave any mention of Franklin's contribution in their acceptance speeches. Only Maurice Wilkins, who received the award jointly with Watson and Crick, praised the 'very valuable contributions' that Rosalind's data had played.[10]

When Franklin's role in the discovery did finally become public knowledge, it was in a way that caused deep distress to her family. In 1968, Watson published his memoir *The Double Helix*, and, although it went on to become a best seller, it was not without its critics. Francis Crick, having read an earlier manuscript, wrote a scathing letter to Watson in 1967, saying that he showed 'such a naïve and egotistical view of the subject as to be scarcely credible' and that Watson's 'view of history is that found in the lower class of women's magazines'.[11] The most controversial aspect of the book was Watson's portrayal of Franklin, or, as he preferred to call her, 'Rosy'—a patronising name by which she was *never* known to family, friends, or colleagues.

Franklin was cast by Watson as a ridiculous figure who alternated between being either a petulant schoolgirl whose 'good brain' was spoiled by an inability to 'keep her emotions under control' or a pantomime villainess who not only obstructed research but also, on one occasion, allegedly even posed a physical threat.[12, 13] Watson seemed to be more concerned with Franklin's lack of lipstick or 'how she would look if she took off her glasses and did something novel with her hair' than with her calibre as a scientist, and his portrayal of Franklin was unrecognisable to her family and friends.[14, 15] When Aaron Klug, a former colleague of Franklin, suggested that at least the book would ensure that Rosalind would always be remembered, her mother simply replied: 'I would rather she were forgotten than remembered in this way.'[16]

Watson's portrayal of Franklin did not go unchallenged. Not long after the publication of *The Double Helix*, those who had known her began to speak out in her defence. With the publication in 1974 of *Rosalind Franklin and DNA* by her friend Anne Sayre,[17] Franklin was adopted as an icon for the growing feminist movement, and hers became a story of how a female scientist suffered in the allegedly oppressive and chauvinistic atmosphere of King's College.[18] For some writers, she came to symbolise the struggle of women in the patriarchal world of scientific research and was even dubbed the 'The Sylvia Plath of molecular biology'.[19] Whether Franklin herself would have been happy to be remembered in this way, however, is doubtful. In a challenge to the appropriation of Franklin as a feminist heroine, the writer Horace Freeland Judson pointed out that to invoke Franklin 'as an example of a woman blocked in her scientific career because she was a woman, diminishes what she accomplished'.[20] It is a view shared by her sister Jenifer Glynn, who, in a recent book *My Sister Rosalind Franklin*, pointed out that Franklin would never have wanted to be remembered as a victim or feminist martyr, but rather for her calibre as a scientist.[21]

Over 50 years later, the controversy over the lack of recognition that Franklin received at the time and her subsequent portrayal by Watson in his book continues. As Nobel Prizes are never awarded posthumously, she would never have been eligible to receive one, but in a recent newspaper interview, Watson was quoted as saying that even had she been eligible, she would never have deserved the Nobel Prize, on the grounds of being 'wrong, stubborn and not getting the answer'.[22] Watson had always maintained that Franklin's failure to spot the significance of 'Photo 51' was because she was 'antihelical' to the point of obstinacy. It is a somewhat unfair judgement. One of Franklin's major contributions was to show that DNA could adopt two different conformations, depending on humidity and water content. At a humidity of less than 75%, the molecule adopted a more compact, tightly coiled configuration, which Franklin called the 'A' form, whereas between 75% and 92% humidity, the molecule took on a longer, thinner conformation. This was the conformation that was thought to resemble the form of DNA most closely in its native cellular environment and Franklin termed it the 'B' form.

The difference was to be highly significant, for each of the two forms gave distinct and different X-ray diffraction patterns. That from the 'A' form had many more distinct spots than that from the 'B' form, and for this reason seemed to be the ideal subject for study by X-ray crystallography as it was believed to contain much more information about the molecule. It was diffraction photos of the 'A' form that Maurice Wilkins presented at the conference in Naples and that fired Watson's ambition.

But one piece of information that these patterns did not suggest was that the molecule was coiled into a helical shape.

The majority of Franklin's work was therefore concerned with an exhaustive analysis of the 'A' form, from which she found no definitive evidence of a helical structure. But far from being 'antihelical', as Watson maintained, the discovery by her former colleague Aaron Klug of a draft paper that she had written shortly before the publication of Watson and Crick's own work in April 1953 revealed that she believed her work on the 'B' form did indeed suggest a helical structure.[23] Having obtained such an important clue, why did she fail where Watson and Crick succeeded?

According to her biographer Brenda Maddox, she had 'come within two steps of answering the most exciting question in post-war science'.[24] The first of these steps was to realise that the two chains of the double helix ran in opposite directions—something Franklin discovered in her work on the 'A' form—but, unlike Watson and Crick, she was not yet in a position to grasp the full interpretation of what it meant. When Franklin disclosed these data in a report to the MRC, which was then shown to Watson and Crick, they quickly grasped its implications for their proposed structure. The second step was to grasp that the double-helical structure allowed the bases on these opposite chains to pair up with each other in very specific ways.[25]

What hindered Franklin was not stubborn opposition to a helical structure, but a conviction that the only time to start building models, as Watson and Crick were doing, was when every last drop of data had been wrung from experimental work. For Franklin to do otherwise would have been what her biographer Brenda Maddox called 'an outrageous leap of the imagination as out of character [for her] as running up an overdraft or wearing a red strapless dress'.[26] For this reason, she focused her efforts on meticulously exhausting the possibilities of the 'A' form before turning her attention to the 'B' form, a point that was emphasised by her sister when she responded to James Watson's comments in his interview with The Times.[27,28]

While the controversies and arguments over Franklin and her photograph continue, at last she is now recognised for her science and its part in the discovery of the double helix. Leading science historian Patricia Fara has said, when discussing the importance of 'Photo 51', that 'this photograph . . . deserves to be celebrated, because it provided clinching evidence for the structure of DNA'.[29,30] With 'Photo 51', Franklin had come close to solving one of the biggest mysteries in biology. Had she observed the X-ray diffraction pattern made by 'B' form DNA even only a year earlier, she might well have had time to address the two issues that still eluded her and gone on to discover the structure herself. It may be tempting to dismiss this suggestion as mere speculation with no grounds in historical reality; however, constructing counterfactual scenarios such as this can be a useful exercise, for they can serve to identify the key factors on which historical events turn—in this case, 'Photo 51' and its distinctive cross pattern. The possibility that Franklin might well have solved the structure of DNA herself thanks to this vital clue carries a particularly intriguing twist. For Franklin wasn't actually the first person to observe this striking pattern at the heart of 'Photo 51' that spoke of a helix and set James Watson's pulse racing. And although the X-ray studies of DNA at King's had first been done by Franklin's colleague, Maurice Wilkins, he was not the first person to study the genetic material using this method. Nor, for that matter, were Watson and Crick the first to attempt to build a model of what the DNA molecule might look like.

Sir Isaac Newton is said to have once remarked that his scientific discoveries were made only thanks to having stood on the shoulders of giants. This is a popular view of the history of science and perhaps a somewhat oversimplified one, but Franklin might well have agreed with Newton.[31] The name of this particular titan was William T. Astbury and he too had been present in the audience at the Naples conference in 1951 when Wilkins presented his X-ray diffraction patterns of DNA that inspired Watson to solve the structure (Figure 4). But whereas Wilkins' pictures of DNA heralded the start of Watson's structural work on DNA, for Astbury they were quite the opposite. Working at the University of Leeds from 1928 until his death in 1961, he had pioneered the use of X-rays to probe the structure of giant biological molecules and, in so doing, laid the foundations for the methods that Wilkins and Franklin would later use to study DNA. In the course of this work, he and his research assistant Florence Bell made both the very first successful X-ray studies of DNA and the first attempt to build a model of the molecule, from which Watson acknowledged that he and Crick later derived key measurements when making the first attempts to build their own model. The climax of this work, however, came in 1951, when Astbury's research assistant Elwyn Beighton obtained a beautiful series of X-ray diffraction photographs of DNA that showed

Figure 4 William Astbury exhibiting a model of a chain molecule. Origi-
nally taken for the *Vogue Book of British Exports*, 1946. Photographer unknown.
In Astbury Papers, MS419 Box A.2, University of Leeds Special Collections,
Brotherton Library.

Reproduced with the kind permission of the University of Leeds, Brotherton Library
Special Collections.

exactly the same black cross pattern that would, two years later, make
James Watson's jaw drop and set his pulse racing.

Yet after attending the meeting at Naples in 1951 where Wilkins presented his work and Watson found his inspiration, Astbury did no more
work on DNA. At first sight, his failure to grasp the vital clue that was
before him seems like a monumental error. But it would be a mistake to
dismiss Astbury as a mere 'also-ran' who failed to cross the finish line in

the race for the double helix. His scientific legacy extended way beyond DNA, and he did far more than just blaze the trail for Watson, Crick, Wilkins, and Franklin. Through his X-ray studies of biological fibres, he founded a whole new scientific discipline, which he popularised as 'molecular biology'—the legacy of which is felt powerfully today. At the heart of this new science lay the idea that biological systems could be understood in terms of giant molecules and how these molecules changed shape. It was this single idea that defined his career, lay at the heart of his success, and, in an ironic twist, may ultimately have cost him the prize of discovering the structure of DNA.

While the names of Watson, Crick, and Franklin are now well known, Astbury has largely been forgotten or, if remembered at all, is dismissed as having 'made serious errors'—a mere footnote in the story of DNA.[32] Yet, what will hopefully become apparent is that, while he may have failed to solve the structure of DNA himself, Watson and Crick might very well have been left emptyhanded without Astbury. In his time, he was a scientist of international stature and one whose work was renowned and respected. The Austrian chemist Max Perutz, who won the 1962 Nobel Prize in Chemistry for his X-ray crystallographic work that solved the structure of the blood protein haemoglobin, once hailed Astbury's laboratory in Leeds as 'the X-ray Vatican'.[33] Following Astbury's death in 1961, his colleague and fellow pioneer in the field of X-ray crystallography, John Desmond Bernal, declared that 'there passed one of the most characteristic figures of what may be called the heroic age of crystal structure analysis . . . but he was more than this . . . he was one of the great characters of British science in the twentieth century'.[34] In the same piece, Bernal went on to say:

> His monument will be found in the whole of molecular biology, a subject which he named and effectively founded. But to those who knew him and had the good fortune to work with him he will always be remembered as Bill Astbury, someone who made you glad to be alive.[35]

Another long-standing colleague, Kenneth Bailey, said:

> The many honors he received were less important to him personally than as a recognition of the new Science of which he was both Master and Prophet, a world of order and plan, the vast field of Molecular Biology as we know it today. He lived to see it flourish and he lived also to see the crowning triumph of X-ray crystallography in the structures of DNA, myoglobin and haemoglobin. To me, a privileged friend for more than

25 years, his work and character are inextricably mixed. Time may blur the edges of his personality but will not obscure the pioneer qualities so evident in his writings that led him forth into his great adventure in the world of fibrous molecules.[36]

Nor was Astbury remembered only for his science. His close colleague at Leeds, Reginald D. Preston, fondly recalled him as being 'a man of many parts—scientist, scholar, musician, bon viveur, humorist, in some ways, a swashbuckler', who was always 'boisterous to the end with every morning still a Christmas morning'.[37,38]

So who exactly was this neglected pioneer whose name has been forgotten for so long, and how did his greatest scientific idea come to be expressed in the form of a rather unusual overcoat?

2

'Germany Has Much to Teach Us . . .'

On Monday 20 July 1981, as the England cricket team faced their old rivals Australia at Headingley, Leeds for the third match in the Ashes test series, their prospects could not have looked bleaker. Having lost the first match of the series at Trent Bridge and drawn the second at Lord's, they were now 227 runs behind and a humiliating defeat looked imminent. The odds of an England victory were being flashed up on the scoreboard at 500–1 and many players were so convinced of inevitable defeat that they had already checked out of their hotels that morning. Then, Ian Botham stepped up to the crease and made cricketing history in one single afternoon. Having just resigned the captaincy, Botham, too, must have had his doubts—if so, it did not show. With a performance that has become legendary, Botham turned the tide of the game and ensured that England not only won the day's play but also went on to win the entire series. Two hundred and forty miles away in London, dealers on the Stock Exchange halted their trades to watch the action at Headingley unfold. Botham, said the *Daily Telegraph* journalist Michael Henderson, 'roared that day and the Aussies cowered . . . That is why Headingley matters.'[1]

But Botham and his fellow cricketing giants are by no means the only reason 'why Headingley matters.' Lying just a couple of miles from Leeds city centre, its Anglo-Saxon past is reputedly reflected in the names of two popular pubs, 'The Original Oak' and 'The Skyrack' (Shire-Oak). Both names are said to refer to an ancient oak at which Saxon elders used to meet; the area is now better known as a watering hole for the city's large student population.[2] With its close proximity to the city's two universities, the area has a reputation for being a quarter for intellectuals and artists and has at various times been the home of writers such as J. R. R. Tolkien and Arthur Ransome, as well as being where the playwright Alan Bennett grew up. It has also been the home of scientific giants, for it was here that William Astbury and his mentor Sir William Bragg both lived.[3]

Astbury's old house stands in a terrace on Kirkstall Lane, directly across the road from the entrance to the test cricket ground where Botham gave his legendary performance in 1981. It was here that Astbury established himself as a scientist of international renown, making the very first studies of the structure of DNA and helping to found a whole new science. Today, these achievements are honoured by a commemorative plaque unveiled by Leeds Civic Trust in 2010, at his former home (Figure 5). Yet while Headingley became his adopted home where he lived for most of his life and found fame and success, Astbury was not a native of Leeds. He was born on 25 February 1898 in the market town of Longton, in the district of Stoke-on-Trent, where his father worked as a

Figure 5 Commemorative plaque unveiled by Leeds Civic Trust on Kirkstall Lane, Leeds, honouring Astbury's life and work.

Photo taken by K. T. Hall.

potter's turner and furniture maker. Known both as the 'Five Towns', or the 'Potteries', the chief industry of the region was ceramics, for which it was nationally famous. Another native of the region was the writer Arnold Bennett, whose work captured the Potteries' life and its inhabitants in vivid detail, and in his novel *The Old Wives' Tale* Bennett highlighted the importance of the industry to the area:

> whenever and wherever in all England a woman washes up, she washes up the product of the district; that whenever and wherever in all England a plate is broken the fracture means new business for the district.[4]

Kenneth Bailey, Astbury's colleague and fellow native of North Staffordshire, once described the Potteries as 'an island jungle of pot banks, kilns, brickyards, and steelworks, surrounded by a pleasant and varied countryside'.[5] An even more vivid and evocative description was offered by Arnold Bennett, who once described Longton, or Longshaw as he called it, as:

> an architecture of ovens and chimneys; for this its atmosphere is as black as its mud; for this it burns and smokes all night, so that Longshaw has been compared to hell; for this it is unlearned in the ways of agriculture, never having seen corn except as packing straw and in quartern loaves; for this, on the other hand, it comprehends the mysterious habits of fire and pure, sterile earth; for this it lives crammed together in slippery streets where the housewife must change white window-curtains at least once a fortnight if she wishes to remain respectable; for this it gets up in the mass at six a.m., winter and summer, and goes to bed when the public-houses close; for this it exists—that you may drink tea out of a teacup and toy with a chop on a plate.[6]

The local inhabitants were 'sturdy and solid with a facet of brittle humour which defies their harsh environment', but, according to Bailey, their lives were constricted by limited opportunities and narrow horizons.[7] 'Fifty years ago', he said, 'few people in the Five Towns were famous in the academic sense, and few had the chance to be'.[8] Yet, of these few, Astbury was one. When his former school of Longton High installed a stained-glass window as a war memorial to honour the 30 former pupils who were killed while serving in the RAF during the Second World War, Astbury's portrait was included. Alongside him were other sons of the Potteries who had gone on to distinguish themselves, such as the physicist Oliver Lodge, the potter Josiah Wedgewood, the aeronautical engineer Reginald Joseph Mitchell (who designed the

Figure 6 A proposal for the instalment of a memorial stained-glass window at Longton High School. Astbury is featured on the top left, above Arnold Bennett.

Reproduced with the kind permission of Mr William Astbury.

Supermarine Spitfire fighter aircraft), and the novelist Arnold Bennett (Figure 6).

According to Kenneth Bailey, Astbury could well have come straight from the pages of one of Bennett's novels. But Astbury's fortunes certainly differed sharply from that of Bennett's most famous character, Anna Tellwright, who, having struggled against a miserly and tyrannical father, remains trapped in a loveless marriage in the Potteries rather than grasp the chance to go to Australia with the man she really loves.[9]

In a eulogy delivered at Astbury's funeral held on 15 June 1961 at Emmanuel Church, Woodhouse Lane, Leeds, Bailey speculated that the motivation for Astbury's drive and determination may well have had its roots in these humble and harsh beginnings:

> I feel perhaps that I had a special affinity with him as originating also, if not within, from somewhere near the five towns. The oppression of those blackened buildings, the pall of smoke, the endless rows of tiny terraced houses might in any sensitive child either defeat or make him. Certainly to Astbury it must have been a great challenge, and in those early days I am sure he felt it was not just important to be a success, but to be foremost in his field of study. This urge explains, I think why he found such inspiration in other works of genius. In music, particularly, genius had a special connotation: it was less amenable to any kind of penetrating analyses than the greatest works of the greatest scientists, and so could always be as wonderful as it was inexplicable.[10]

Perhaps this was also the motivation that drove Astbury's mother to encourage and support her son. Recognising from an early age that he excelled academically, she nurtured his talents, and her drive and determination were rewarded when he won a scholarship to Longton High School. Here his interests included cricket, amateur dramatics, short-story writing, sketching, and a love of music that he inherited from his father, and which would remain with him throughout his life. Together with his younger brother Norman, who later became Director of the British Ceramic Research Association, Astbury would play the works of his musical heroes, who included Bach, Mozart, Beethoven, and Schubert, and he loved to argue that no other composer approached their genius.

It was also while at Longton High that he began to develop his love of science, encouraged by the Headmaster and Second Master. Towards the end of his time there, he became Head Boy and won the Duke of Sutherland Gold Medal, but his crowning achievement was to win the only available scholarship to Jesus College, Cambridge. Astbury went up to Cambridge in 1917, but after only two terms was called up for military service, having enlisted the previous year. However, on account of a poor medical rating following an appendectomy in 1915, he was deemed unfit for combat, and so, rather than serve in France, he was posted to Cork in Southern Ireland, where he served in the Royal Army Medical Corps (RAMC) (Figure 7).[11]

Figure 7 Astbury serving in the Royal Army Medical Corps, Cork, 1917.
Reproduced with the kind permission of Mr William Astbury.

Ireland might have been far from the fighting on the Western Front, but it was not without its own problems. At Easter the previous year, Irish Nationalists had led an insurrection against the British armed with weapons and support from Germany. The Easter Rising, as it became known, left the local population with little sympathy for the British army and it was against this background of hostility that Astbury met a young Irish girl, Frances Gould, who later became his wife and with whom he had two children, William and Maureen (Figure 8). In the wake of the Easter Rising, Frances' romance with a British soldier was not welcomed by certain members of the local community and, according to family sources, met with outright hostility. As a result, Frances left Ireland with few regrets to start a new life in England with Astbury after the war.[12]

(a)   (b)

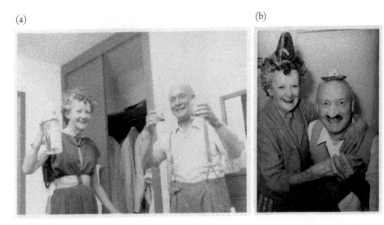

Figure 8 William Astbury—'man of many parts: scientist, scholar, musician, *bon viveur*, humorist'—and his wife Frances Gould enjoying married life.
Reproduced with kind permission of Mr William Astbury.

As well as meeting Frances, Astbury's brief time in Ireland was eventful for another reason. It was during his time in Cork that he first began working with a technology that would come to be the foundation of his scientific career—X-rays. This enigmatic and powerful form of radiation had been discovered in 1895 by the German physicist Wilhelm Röntgen. In his laboratory in Würzburg, Röntgen had been studying the effects of passing electricity through a gas contained in a glass vessel at low pressure. With this work, he was continuing in a long line of research that dated as far back as a series of experiments done by Michael Faraday in 1837. Since then, several other workers, including the English chemist William Crookes, had found that when a high voltage was applied across a gas contained in a tube at very low pressure, the tube emitted an eerie glow. For Crookes, this glow was evidence of a transcendental spiritual world, but it also brought a very practical irritation as it meant that piles of photographic plates in his laboratory became fogged. Röntgen encountered a similar problem, and reasoning that the plates were becoming fogged owing to the light emitted from the discharge tube, he covered the glass vessel in black card. To his surprise, however, he found that his photographic plates still became fogged and screens coated with barium platinocyanide, a fluorescent material, began to glow.

Röntgen's conclusion was that the discharge tube was emitting not only visible light, but also some other kind of previously unidentified radiation that, unlike light, could pass through the black cardboard. Other recently discovered forms of radiation, such as cathode rays and alpha particles, had been shown to be particles of matter carrying an electrical charge, which meant that they could be deflected by a magnetic field.

Röntgen's rays, however, were unaffected by magnetic fields, suggesting that they might be a form of electromagnetic wave like visible light or gamma rays. Unlike visible light, however, Röntgen found that they could not be reflected, nor could they be bent, or 'refracted', as happens when light passes through a prism. But most remarkable of all was their ability to penetrate solid matter. In stark contrast to visible light, they could pass with ease through a range of physical materials: tinfoil, books, wood, and packs of playing cards could not block the new rays. Only dense materials like lead or bone blocked their path, as Röntgen discovered when he held a sheet of paper covered with barium platinocyanide near the discharge tube and saw the shadow of his own hand appear on the glowing screen.

For the next eight weeks, Röntgen ate and slept in his laboratory in an exhaustive attempt to characterise the properties of these strange new rays. During this time, his wife hardly saw him, and when she did, he was tired and irritable—a result no doubt of his frustration at being unable to determine whether the rays were particles like alpha radiation or electromagnetic waves like visible light. Nevertheless, his scientific colleagues were impressed with his work and when he presented his initial findings to the Physikalisch-medicinischen Gesellschaft of Würzburg on 28 December 1895, the work was published immediately in German and an English translation appeared in the journal *Nature* on 23 January 1896.[13]

In honour of their discoverer, the new rays became known in Germany as 'Röntgenstrahlung'; elsewhere, they became known by the more familiar name of 'X-rays'—a term that reflected their enigmatic nature. While physicists remained baffled as to whether Röntgen's rays were particles or electromagnetic waves, the practical implications of their phenomenal power to penetrate solid matter were quickly seized upon. One particularly memorable example of this power was an X-ray image that Röntgen took of the shadow cast by the bones in his wife's hand, and the medical profession were quick to seize upon the possible practical applications that X-rays offered for study of the body.[14] Reports soon began to be published describing how X-ray photography had been used to study kidney stones, bone lesions, and rheumatic joints. A paper published in *The Lancet* by the physicist Oliver Lodge (himself a former native of Stoke who was also honoured in the Longton High memorial window along with Astbury) described how X-rays had been used to locate a bullet lodged inside a limb, raising the possibility that X-rays could be ideal for the treatment of battlefield injuries.[15] As a result, in its 1896 expedition to the Nile, the British Army had a designated X-ray unit and when the RAMC was founded in 1898, such X-ray units were a core feature of its facilities.[16] It was in one such unit that Astbury found himself working as an acting, unpaid lance-corporal in charge of medical X-ray work. Here, his initiative at making improvised repairs to a primitive X-ray machine with some wire and a tin can was not appreciated by his senior officers, nor was the loss of a large number of medical records when a shelf stacked with X-ray plates collapsed. Later in life, he liked to tell how, thanks to these two mishaps, he had set a record for being court-martialled twice during his brief time in the army and would roar with laughter as he recalled how both cases were dismissed as involving 'An Act of God or the King's Enemies'.[17]

X-rays may have blighted Astbury's short military career, but they would prove central to his success as a scientist, and this was all thanks largely to the influence of one man—the physicist Sir William Bragg, whom Astbury later fondly remembered as a scientific father figure. While Astbury may once have described himself as being 'the alpha and omega' of using X-rays to study biological fibres, the roots of his work were to be found in the ground-breaking innovation of William Henry Bragg and his son William Lawrence, and it is with them that the story of his science really began.[18]

While doctors and soldiers used X-rays to study broken limbs and locate bullets, Bragg and his son Lawrence pioneered their use to peer into an even deeper world hidden from the human eye—the atomic structure of crystals. Like Astbury, William Bragg came from a working-class background. He was born near Wigton in Cumbria in 1862, where his father, a former merchant seaman, had purchased a small farm. After the death of his mother, he was brought up by his uncle in Market Harborough, Leicestershire, before attending King William's College on the Isle of Man and then Trinity College, Cambridge, where he read mathematics. On graduation, he took up a post as Professor of Mathematics and Physics at the University of Adelaide, Australia. Although he excelled at mathematics, he knew very little about physics and admitted to having taught himself from several textbooks on the long sea voyage to Australia.[19]

Bragg described his stay in Adelaide as being 'like sunshine and fresh invigorating air', and it was a time of great success for him in both his professional and personal life.[20] It was here that he married Gwendoline Todd, the daughter of the Government Astronomer Sir Charles Todd, and together they had three children, William Lawrence, Robert, and Gwen. It was also while in Adelaide that he honed two key skills that would prove to be crucial to his career. The first of these was in the design and construction of scientific instruments. The University of Adelaide had only recently been founded, and, as a result, Bragg found on his arrival that it was severely lacking in resources. On one occasion, he had to write to the university authorities pointing out that in the mathematical lecture room there were no desks or tables on which students could take notes during lectures.[21] Experimental equipment for the teaching of physics was also in short supply, and, to remedy this situation, Bragg apprenticed himself to a local firm of instrument manufacturers to learn the skills required to build his own equipment. It was time well spent, for it would later pay great dividends.

The other great skill that Bragg acquired while in Adelaide was as a lecturer and communicator of science. Although at first he found lecturing difficult, he soon developed the skill of engaging his students and became well known for his ability to communicate science to both undergraduates and a wider audience at public lectures given in the evening. It was at one such public lecture in June 1896 that Bragg first addressed a subject that would have a profound effect in shaping the subsequent direction of his career.

Bragg later said that 'no scientific discovery before or since that of Röntgen in 1895 has excited such immediate or universal interest'.[22] This immediate excitement was certainly palpable in Adelaide. On the evening of Friday 29 May 1896, Bragg demonstrated to a gathering of local doctors this impressive ability of X-rays to reveal structures that were otherwise invisible. A local chemist by the name of Samuel Barbour, who worked for Faulding & Co., an Adelaide manufacturer of pharmaceuticals, provided Bragg with a glass discharge tube that he had obtained while recently on holiday in Europe. By attaching this tube to a battery borrowed from Bragg's father-in-law and an induction coil from which an initial electrical spark could be generated, Bragg and Barbour were able to generate short bursts of X-rays that left their audience 'highly delighted with the achievements'.[23] Like Röntgen, Bragg was not averse to using himself as an experimental test subject, and one of his X-ray photographs taken that evening showed the bones in his own hand, revealing an old injury to one of his fingers sustained when using the turnip chopping machine on his father's farm in Cumbria.[24]

A week later, Bragg gave a similar demonstration to the Governor of South Australia's wife, Lady Victoria Buxton, who was so impressed that she returned two weeks later with her husband and the Chief Justice.[25] By now, so much excitement had been generated in the local media that Bragg's lecture of 17 June, originally scheduled to be held in the University Library, had to be relocated to the much larger University Hall, where, according to local newspaper reports from the time, many of the audience had to scramble for seats or be turned away at the door.[26] Here, Bragg enthralled his audience by using X-rays to reveal that a box on the table before him contained a number of items, including a pair of spectacles belonging to the Governor's wife, as well as locating a set of keys hidden beneath the table.

Nor was Bragg's fascination with X-rays confined only to public demonstrations of their use. When his eldest son Lawrence fell from his tricycle following an altercation with his younger brother Bob and fractured his arm, Bragg used a home-built X-ray emission tube to study the injury. Years later, Lawrence recalled his terror at being one of the first people to undergo a medical X-ray examination, saying:

> I was scared stiff by the fizzing sparks and smell of ozone and could only be persuaded to submit to the exposure after my much calmer small brother Bob had his radiograph taken to set me an example.[27]

Bragg's public lectures may well have been the beginning of his life-long interest in X-rays, but it was for other work that he would first achieve fame among international circles of scientists. When asked to give a presidential address to the Australian Association for the Advancement of Science, recent work done by Marie and Pierre Curie on radioactivity caught his attention. Aged 42, Bragg had until now done no scientific research, concentrating instead on teaching and educational reform. Such a situation would be unthinkable in contemporary academia and would certainly sound the death knell for any hope of a scientific career. But, inspired by the Curies' work, on Saturday 30 July 1904, in a small basement on the university campus, Bragg began his first experiments on the type of radioactive emission known as alpha rays.[28] His subsequent paper reporting on the penetrating power of these particles emitted from unstable nuclei of the element radium was hailed as a major contribution to the study of this new field and established Bragg as a physicist of international renown. As a consequence of this new work, he established a regular correspondence with the physicist Ernest Rutherford at McGill University, Canada, who was one of the major pioneers in the study of radioactivity and atomic structure, and it was with the support of Rutherford that in 1907 Bragg achieved one of the greatest honours in his career when he was elected to be a Fellow of the Royal Society.

But Bragg's new professional standing also brought new challenges. Elevation to international fame meant that he now had to defend his research against criticism from rivals. In the same year that he was elected to the Royal Society, Bragg began a long-standing correspondence with one particularly vocal critic, the English physicist Charles Barkla at the University of Liverpool. One feature of alpha particles that intrigued Bragg was their ability to strip electrons away from atoms of gas leaving the gas electrically charged, or ionised. That X-rays possessed this same ability was for Bragg strong evidence that, like alpha rays, X-rays might also be composed of a stream of material particles as opposed to being electromagnetic waves. Unlike alpha particles, however, which carried a positive electrical charge and so could be deflected by a magnetic field, X-rays were unaffected by magnetism. Bragg proposed that this could be explained if the X-rays were composed of pairs of particles carrying opposite electrical charges, or what he termed a 'neutral pair'. But the opinion of physicists about the nature of X-rays remained divided, and one of the chief proponents of the wave nature of X-rays was Barkla. In 1907, Barkla published a paper showing that, like visible light, X-rays could be polarised—a property that could only be explained if they were waves and not particles. It was a formidable challenge to all those who advocated a particle model of X-rays, and Bragg quickly found himself having to respond to Barkla's work. In the years that followed, Bragg and Barkla aired their disagreements publicly in the pages of the journal *Nature*. What was particularly notable about these exchanges is that, at times, Bragg's usual tone of gentle reservation slipped away to reveal a certain uncharacteristic irritation and annoyance.

What became known as the 'Bragg–Barkla controversy' has been discussed in detail by historians of science and is of particular interest here for what the affair reveals about Bragg's stance towards experimental science. Having built his career so far on improving teaching and educational reform at Adelaide, with the success of his work on alpha particles, Bragg now wanted to concentrate on experimental research and felt that in order to do this it would be necessary to leave Adelaide. When Rutherford accepted a post in Manchester, England, and his post at McGill was vacant, he suggested that Bragg would be a fine candidate to replace him. But a fire at McGill left the University's finances devastated and consequently they were no longer able to offer Bragg the position. Thankfully, however, he had another option. On the other side of the Pennine hills, the University of Leeds hoped to emulate Manchester and recruit a physicist of similar calibre and international standing to Rutherford.

In 1909, Bragg arrived in Leeds to take up the post of Cavendish Professor of Physics, but his connection with the city had already been established while he was still in Adelaide. Barbour, the local chemist who had helped Bragg at the evening lectures in Adelaide, was a photography enthusiast and, while on holiday in Europe, had visited his home city of Leeds. By this time, Leeds had become an active centre for the development and pursuit of new methods in photography. Only a few years earlier in 1885, Louis Aimé Augustin le Prince, a Frenchman living in the city, had patented the first motion picture film process and shot some of the earliest moving film footage of traffic and pedestrians crossing Leeds Bridge.[29] It was while on this visit to Leeds that Barbour acquired two glass discharge tubes from Reynolds & Branson, a company based on Briggate in the city centre, that specialised in the manufacture of laboratory and photographic equipment and of which one of the founders, Frederick Branson, was an important local figure in promoting photography and the new technology of X-rays.[30–32] On his return to Adelaide, Barbour gave the discharge tube to Bragg for use in his evening lectures, and it was with this that he generated the X-rays that thrilled and captivated his audiences.

Bragg's first years in Leeds, however, were a stark contrast to his previous life in Adelaide. Whereas he had likened Adelaide to sunshine and fresh air, he said of his new home that 'Leeds people are really very nice: all those I met anyway. The place itself is grimy, even the suburbs; but you can get out into beautiful country to the North . . . '.[33]

The grime of which Bragg spoke resulted from the city's prominence as a centre of industry and manufacturing. Though mentioned as 'Loidis' in the ancient British kingdom of Elmet, by the Venerable Bede in his chronicle *Ecclesiastical History of the English People*, the first detailed account of the town and its surroundings appears in the Domesday Book of 1086, in which the Manor of Leeds, with an estimated population of 200 and a value of £7, fared much better than the outlying settlements such as Headingley, whose values had suffered as a result of the brutal campaign known as 'The Harrying of the North', carried out by William the Conqueror to crush northern resistance to his rule.[34–36] By the time of Bragg's arrival, Leeds had undergone massive urban growth to become a major international centre of manufacturing, having been one of the first cities to experience rapid industrialisation during the end of the eighteenth and the start of the nineteenth century. The effect on the local landscape and character was profound for, as Prince Herman von Pückler, visiting from Germany, remarked:

I reached the great manufacturing town of Leeds just in the twilight. A transparent cloud of smoke was diffused over the whole space which it occupied, on and between several hills; a hundred red fires shot upwards into the sky, and as many towering chimneys poured forth columns of black smoke. The huge manufactories, five storeys high, had a grand and striking effect. Here the toiling artisan labours far into the night . . ..[37]

But while the German prince was left in awe by the sheer scale of the factories, mills and chimneys, others were less than impressed, as the following verse, which has been ascribed to the writer William Hazlitt, testifies:

O smoky city, dull and dirty Leeds,
Thou may'st be well for trade and eke for wealth,
And thou may'st cleanse thyself at times by stealth,
Like men who do, but never own, good deeds[38]

As a result of these developments, by the 1890s, Leeds was home to a diverse number of industries, including printing, paper production, dyeing, glass manufacture, pharmaceuticals, brewing, and engineering. Thanks to the work of pioneers in engineering such as Matthew Murray, Charles Todd, and James Kitson, the city became a centre of the railway industry, and locomotives made in the foundries of Hunslet were exported around the world.[39] In 1893, Leeds was hailed as:

A town of the times ... whose labours are for the welfare of mankind, and whose products have the whole world for their market ... Though Leeds may lack the charm of Greece and Italy ... she can place in the counterbalance her nine hundred factories and workshops, monuments of her wealth, industry and prestige.[40]

This sense of provincial Victorian civic pride was also evident in the magnificent public buildings designed by the architect Cuthbert Brodrick, such as the Town Hall, the Corn Exchange, and the Mechanics Institute (now home to the City Museum), which still stand in the city centre and have lost none of their majesty.

But there was one industry in particular to which Leeds owed its wealth and one that would play an important role for both Bragg and, later, Astbury. Ever since the Cistercian monks at Kirkstall Abbey, founded in 1152, had supported themselves by raising sheep on the lush green banks of the Aire valley and selling their fleeces, wool had been at the heart of the economy of Leeds, and this is still reflected in the city's coat of arms, which depicts a hanging fleece.[41-44] Nestled in the foothills of the Pennines, the local geology gave Leeds a particularly favourable setting for the production and processing of wool. To the north of the city were high areas of millstone grit, a type of sandstone, and beneath the city were alternate layers of sandstone and shale at shallow depths, which provided naturally filtered water.[45] The result was that the streams running from the north of the city contained a very soft water ideal for washing and scouring wool, which was then exported down the River Aire and via the ports at Scarborough and Hull to merchants from Italy, France, and the Low Countries.

The years between 1450 and 1550 saw rapid growth of the local textile industry, and by the seventeenth century, despite the disruption caused by fighting during the English Civil War in which the Parliamentarians eventually won control of the town, the wealth of Leeds had become founded on wool.[46-48] The growth in the industry gathered new momentum with the introduction of mechanisation in the eighteenth century. The fast-flowing streams north of the city were able to power water mills and the vast coalfields to the south provided the fuel for steam power. On visiting Leeds in the 1720s, the writer Daniel Defoe was notably impressed by the size and scale of the city's textile industry, saying of the local cloth market that it was 'a Prodigy of its Kind and is not to be equalled in the World'.[49] The cloth merchants of Leeds seemed to have a particular talent for identifying new markets, such as the 13 American colonies, which, together with the West Indies, received 30% of English woollen exports, and, by 1770, Leeds merchants accounted for one-third of all wool exports from England, with the value of cloth exports from Leeds amounting to £1,500,000.[50,51] Despite suffering a depression during the American War of Independence, the industry once again enjoyed a boom when hostilities finally ended in 1783 and, by 1800, the Leeds cloth merchant Benjamin Gott was the twelfth-largest employer in the country thanks to his fulling mill in Armley, which was believed to be the largest in the country at the time and is now an industrial museum.[52,53]

By the time of Bragg's arrival, Leeds was a wealthy city, thanks in no small part to the growth of its textile industry. But, according to his daughter Gwen, Bragg was left saddened by the gulf between the richness on display in certain quarters of the city and the abject poverty in others.[54] In addition, if he had hoped that this wealth might provide first-class conditions on which to embark upon his planned research, he was disappointed. Instead of finding state-of-the-art laboratories at his disposal, Bragg found himself housed in temporary shedding that was freezing during the winter and was not replaced until 1932. In a letter to Rogers, his instrument maker in Adelaide, he showed characteristic restraint in hiding his disappointment, tactfully describing his new working conditions as 'curious'.[55] Nor was this the only challenge in his professional life, for now the power to inspire his students in lectures seemed also to have deserted him. According to his daughter, 'In Adelaide, students had drunk in his words; the Leeds students he failed to win—they stamped their feet; his lecturing went badly.'[56]

The move from Adelaide also came as a culture shock to his wife Gwendoline, who was:

> appalled by Leeds, by the dirt, by the smoky dark, by the rows of little poor back-to-back houses. The only bright spots on the scene were the whitened doorsteps, which one must step *over*, not *on*. She was horrified by the pasty-faced babies carried folded into their mother's shawls, and the rickety children.[57]

Gwendoline's philosophy, however, was that the best way to overcome her initial dislike of her new home was to become actively engaged with the life of the city. When the family moved from rented accommodation to their own home, Rosehurst on Grosvenor Road in

Headingley, she resolved to make their time in Leeds a success. Now largely the student quarter of the city, Headingley was in those days the home of the industrialists who had made Leeds wealthy, and it was with their families that Gwendoline began to cultivate a busy social life. She was elected to the 'Little Owls', a select ladies circle, and the Art Club, as well as becoming actively involved in social work for the poor of the city. Although the inequalities in the distribution of wealth in Leeds made William uncomfortable, he was also heartened that 'this is a surprising city for charity and good works, and everybody is expected to do something'.[58] Far from wallowing in regret at having left Australia, Gwendoline did indeed 'do something' by devoting her time and energy to the 'Babies Welcome', a charity dedicated to the care of undernourished babies and children.

Gwendoline's efforts helped to make the move to Leeds more bearable, but the family nevertheless felt the need to escape the grime of the city and the pressures of work. In their second year at Leeds, they found solace by renting a small farmhouse called 'Deerstones' near Bolton Abbey in Lower Wharfedale. Here they made regular retreats to spend time painting amidst the inspiring scenery of the Yorkshire Dales, and Bragg's daughter Gwen recalled their happy times there:

> I loved Deerstones passionately, and I saw the grown-ups happy there too. WHB [*William Henry Bragg*] enjoyed the simplicity and the quiet; it had a special peace filled with the rustle of beech trees and the sound of the beck; and GB [*Gwendoline Bragg*] baked her own bread and painted in watercolour. My brothers went sketching too, and friends came out from Leeds.[59]

Retreats into the Yorkshire countryside provided welcome respite from the stresses of his professional life, but with Bragg's fame and reputation, it was never possible to escape completely. One of the biggest challenges in his professional life was his ongoing disagreement with Barkla over the nature of X-rays. The debate had in fact become so heated that the editor of *Nature* had declared that the correspondence should now be concluded. Whereas Bragg continued to support the idea of X-rays being made of particles, Barkla continued to gather evidence that they were, in fact, electromagnetic waves. It was while on holiday at Cloughton, near Scarborough, on the East Yorkshire Coast during the summer of 1912 that Bragg received a letter informing him of a recent experiment in Germany that appeared to give considerable support to Barkla. One might expect this news to have been met with disappointment, but in fact the letter proved to be a pivotal moment in Bragg's career, for it was in response to this news that he began the work that would eventually lead to the Nobel Prize.

The letter was written by a former research student called Lars Vegard who had worked with Bragg in Leeds before his return to Norway and it contained the details of some experimental work done at the University of Munich by three German scientists: Max von Laue, Walther Friedrich, and Paul Knipping. Like Bragg and Barkla, von Laue was intrigued by the question of whether X-rays were particles or waves and reasoned that, if they were waves, they ought to display the phenomenon known as diffraction. This occurs whenever a wave passes through a gap of comparable size to the distance between successive peaks or troughs in the wave motion (i.e., the wavelength), resulting in a spreading of the wavefront. It can be achieved with visible light by passing light of a single wavelength through a pair of double slits known as a diffraction grating, resulting in an alternating pattern of light and dark bands on a screen. Von Laue reasoned that if X-rays were waves, then their wavelength must be of the order of only a few nanometres and what puzzled him was where to find a gap of a similar size that would result in their diffraction. After reading the thesis of PhD student Paul Ewald, who had come to consult him over a question regarding optics, von Laue found an answer. From Ewald's thesis, he calculated that the spacing between individual atoms in a crystal must be of the same order as the hypothetical wavelength of X-rays, and if so, then the crystal itself could act as a diffraction grating for an X-ray beam, with the resulting diffracted X-rays forming an interference pattern of black spots on a photographic film.

At one of their regular gatherings at the Café Lutz in the Hofgarten (later to be immortalised in T. S. Eliot's poem, 'The Wasteland'), von Laue presented his ideas to his colleagues. Arnold Sommerfeld, the Professor of Theoretical Physics, was unconvinced, pointing out that any scattered X-rays would be so buffeted by the thermal vibrations of the atoms in the crystal that they could never generate any meaningful patterns. But some of the younger members of the group were more optimistic, and von Laue was able to persuade Friedrich, Sommerfeld's newly appointed experimental assistant, together with Paul Knipping, a former student of Röntgen, to test his hypothesis experimentally.

By passing X-rays through a fine 3-mm hole into a box with lead walls, Friedrich and Knipping created a tightly focused beam that then fell upon a crystal of copper sulphate mounted in the centre of the box. Just as von Laue had predicted, the emerging X-rays were indeed diffracted, and the interference patterns resulting from the scattered rays could be observed as an arrangement of discrete dark spots on a photographic film. When the experiment was repeated with crystals of zinc sulphide, or 'zincblende', the results were even better, with clearer, sharper spots on the photographs, and on 8 June and 6 July 1912, the work was presented to the Bavarian Academy of Sciences.

The diffraction experiments by von Laue, Friedrich, and Knipping were strong evidence that, like visible light, X-rays were another form of electromagnetic wave but vibrating at a higher frequency. To borrow a musical analogy, it was as if a whole new higher octave of notes had been discovered. It was also a formidable challenge to all those physicists like Bragg who argued that X-rays were composed of a stream of particles. Yet rather than leave Bragg feeling deflated, Vegard's letter spurred him and Lawrence into action.

Lawrence in particular was intrigued by the theory behind von Laue's work, and it was with this that he wrestled on his return to Cambridge after the summer holiday. As an undergraduate at Trinity College, he had initially read mathematics but later transferred to study physics, and both these academic disciplines now proved invaluable to him in the interpretation of von Laue's results. In an effort to provide a theoretical account of his results, von Laue had attempted a complex mathematical explanation that treated the crystal as a three-dimensional diffraction grating. But this approach had brought him into severe difficulties. Not only was the mathematics becoming increasingly complex, but von Laue was working with incorrect dimensions for his crystal of zincblende and also mistakenly thought that the scattered X-rays forming black spots on the photographic film were not from the original X-ray beam but resulted rather from a secondary pulse of X-rays generated by the crystal itself. It was while strolling along 'The Backs' in Cambridge that Lawrence found inspiration in a much simpler approach. Thinking back to his lectures in optics, he imagined that the regular rows and columns of atoms and molecules in the crystal could be thought of as individual planes of reflecting mirrors. Each of these 'mirrors' was inclined at a different angle and bounced the passing X-rays in different directions. Each spot on von Laue's photographic plates could then be thought of as resulting from X-rays reflected by a specific layer of atoms at a particular angle. By reformulating the problem in this way, Lawrence arrived at a simple equation. Known as Bragg's law, this equation established a clear relationship between the spatial arrangement of atoms in the crystal, the angle of a scattered X-ray beam, and the wavelength of the X-rays; the equation has carried his name into the pages of A-level physics textbooks ever since.

In November 1912, Lawrence Bragg published his ideas about von Laue's pictures in a paper entitled 'The diffraction of short electromagnetic waves by a crystal' in *Proceedings of the Cambridge Philosophical Society*. The tone of the paper was one of guarded speculation, for he was anxious not to completely contradict his father's view of X-rays as particles.[60] But, far from feeling threatened, William was excited by his son's work and wrote to Rutherford in Manchester, exclaiming: 'My boy has been getting beautiful X-ray reflections from mica sheets, just as simple as the reflections of light in a mirror.'[61] Lawrence was keen to develop this work further, but found his efforts frustrated by difficulties in his working environment. Often, he was let down by poor and unreliable equipment that he had had to cobble together with drawing pins and cardboard. Writing many years later, Lawrence recalled the difficulties of his situation:

> When I achieved the first X-ray reflections I worked the Rumkorff coil too hard in my excitement and burnt out the platinum contact. Lincoln, the mechanic, was very annoyed as a contact cost ten shillings [*a week's wages at the time*], and refused to provide me with another for a month. I could never have exploited my ideas about X-ray diffraction under such conditions.[62]

His father up in Leeds, however, faced no such problems. Inspired by Vegard's letter and Lawrence's initial results, William had spent the latter half of 1912 putting the practical skills that he learned in the manufacture of scientific instruments while in Adelaide to good use. Working with Mr Jenkinson, a skilled mechanic, at Leeds, William built the first X-ray spectrometer—a device that allowed the spectra of the diffracted X-rays to be studied. The advantage of the spectrometer was that it enabled a far more powerful analysis of X-ray scattering from crystals than von Laue's photographic method (Figure 9). X-rays of a single wavelength were generated in a discharge tube and then directed into a chamber where the crystal for examination was mounted on

Figure 9 William Bragg's X-ray spectrometer.
Reproduced with the kind permission of the University of Leeds, School of Physics and Astronomy.

a rotating platform. By turning the platform, the angle between the incident X-ray beam and the crystal could be varied. The angle and intensity of the resulting scattered X-rays could then be measured as they passed into another chamber, which was filled with gas. As the X-rays passed through the gas, they stripped off electrons from the gas molecules, leaving them ionised and so giving rise to an electrical current that was directly proportional to the intensity of the scattered X-ray beam. The advantage of using the X-ray spectrometer over photographic film was that it was a much less laborious and more direct way of measuring the scattered X-rays than trying to decipher the numerous spots on von Laue's pictures. Whereas the spots on a von Laue photograph could be thought of as representing the sum total of all the reflections from every different plane of atoms in the crystal, the X-ray spectrometer enabled the scattering from each individual reflecting plane to be analysed one at a time. In addition, the X-ray spectrometer also gave quantitative information on the scattered X-rays.

When Lawrence came home from Cambridge at the end of 1912, he and his father spent the Christmas holidays working feverishly together. William's main interest was to use the spectrometer to determine the properties such as the wavelength of the different kinds of X-ray scattered by the crystal. Lawrence, however, was far more interested in what the scattered X-rays might reveal about the crystal itself, for it had struck him that the particular angles at which X-rays were scattered might be a clue to the specific arrangement of atoms in the crystal. And he reasoned that, by working backwards from measurements of the scattered X-rays, one ought to be able to deduce the positions of the atoms in the crystal.

By combining Lawrence's theoretical ideas and William's instrumentation to test them, the father and son team published their first joint paper, entitled 'The reflection of X-rays by crystals'. Published in 1913 in the *Proceedings of the Royal Society of London*, it established the basic principles of a new scientific technique that became known as X-ray crystallography.[63] In time, this ground-breaking method first developed by William and Lawrence Bragg would come to be applied across a broad range of scientific disciplines, including geology, chemistry, and materials science. But the one scientific discipline to which X-ray crystallography would become particularly invaluable was biomedical research, where not only would it become crucial to solving the structure of life-saving drugs such as insulin and penicillin, but it would also be the method that William Astbury and Florence Bell used to make the first studies of the DNA molecule and with which Rosalind Franklin would eventually take 'Photo 51'.

Throughout 1913, William and Lawrence continued to publish a steady stream of papers, either as single or joint authors, showing how their new method could be used to study the arrangements of atoms in inorganic crystal samples that Lawrence had smuggled up to Leeds despite the Professor of Mineralogy at Cambridge issuing a strict edict that no crystal samples should leave his department. After the summer, instead of returning to Cambridge, Lawrence spent the whole of the Autumn term working in Leeds with his father. Straight after breakfast, father and son would walk together from their house in Headingley past Hyde Park Corner and along Woodhouse Lane to reach William's laboratories in the 'Physics sheds'.[64] It was a turning point in William's relationship with Leeds and, as his daughter later said, 'the sun shone for him as it had not since Adelaide'.[65] Recalling the excitement of this time, Lawrence said:

> The X-ray spectrometer opened up a new world. It proved to be a far more powerful method of analysing crystal structure than the Laue photographs which I had used. One could examine the various faces of a crystal in succession, and by noting the angles at which and the intensity with which they reflected the X-rays, one could deduce the way in which the atoms were arranged in sheets parallel to these faces. The intersections of these sheets pinned down the positions of the atoms in space. It was like discovering an alluvial gold field with nuggets lying around waiting to be picked up. It was a glorious time when we worked far into every night with new worlds unfolding before us in the silent laboratory.[66]

At the same time, however, Lawrence's delight was tempered with a certain frustration. Although he had worked out the theory underlying X-ray diffraction while his father had developed the experimental apparatus to carry out the work, it was William Bragg, as the established senior figure of the partnership, who presented their research at the British Association Meeting and the Solvay Conference on the Structure of Matter in 1913. Although William made a point of stressing that it was his son who had worked out the theory underlying X-ray diffraction, it was perceived by many of his peers as having been William Bragg's own innovation. It was a difficult situation, and one that left Lawrence feeling somewhat hurt.[67] When, many years later, as the head of the MRC Laboratory in Cambridge, he was shown 'Photo 51' by James Watson, it was perhaps the memory of this experience and an unwillingness to lose out once again that made him allow Watson and Crick to resume their work on DNA.

By 1914, Lawrence had his own X-ray spectrometer in Cambridge and was making further studies into the atomic structure of inorganic crystals such as zincblende, fluorspar, calcite, and iron pyrites. Meanwhile, his father in Leeds concentrated on using the spectrometer to probe further into the nature of X-rays and study the structures of diamond and ice. It was a golden age for the father and son team, during which they produced a flurry of scientific papers. Sadly, it was not to last. Immersed in recording measurements on their spectrometers, it is unlikely that the Braggs paid much attention to news from a remote corner of the Balkans. But the assassination in Sarajevo of Franz Ferdinand, Archduke of the Austro-Hungarian Empire, was to prove the spark that would set the whole of Europe alight. For several decades, a precarious balance of power had been maintained in Europe through a series of treaties and alliances. In the wake of the assassination in the Balkans, the fragile peace began to unravel with alarming speed, and in August 1914, under obligation by a treaty to protect Belgium, Britain declared war on Germany.

In July of that year, Lawrence had asked his father about the possibility of attending a scientific conference in Germany; by August, he was serving as a second lieutenant in the Leicestershire Royal Horse Artillery. Here his skills as a physicist served him well. The French Army had been experimenting with a novel method of locating the position of enemy guns based on the sound of their firing, and, on account of his academic background, Lawrence was thought to be the ideal person to help with this work. In August 1915, he was posted to the Western Front, where he helped to set up the very first sound-ranging station, an achievement for which he was awarded the Military Cross and earned three mentions in Dispatches. But this was not the only honour that was bestowed upon him during this time. It was while stationed south of Ypres close to the front line that he received a communication informing him that he and his father had jointly been awarded the Nobel Prize in Physics for their development of X-ray crystallography—an achievement that was celebrated by the local village priest opening a bottle of Lacryma Christi. To this day, William and Lawrence Bragg are the only parent–child team ever to be jointly awarded a Nobel Prize.

Winning the Nobel Prize should have been the crowning moment of the Braggs' scientific career, but their memories of 1915 would be forever blighted by personal tragedy. With the outbreak of war, William's youngest son Robert, known as Bob, had joined the Royal Field Artillery and been posted as part of the British Mediterranean Expeditionary Force to take part in the Allied landings at Gallipoli. The hope was that by capturing and taking control of the Gallipoli Peninsula, the Ottoman Empire would be forced to surrender and a new supply line to Russia could be opened up via the Bosphorus. Unfortunately, the plan failed, and the attack ended in disaster, with the Allies suffering heavy casualties, including Bob. Shortly after landing at Gallipoli, a dugout in which he had taken shelter was hit by a Turkish shell. Having lost a leg and sustained severe injuries to the other, Bob was hurried aboard a hospital ship but died from his wounds the next day.

Having been informed of Bob's death by a telegram that arrived in Leeds, William joined his wife and daughter up at 'Deerstones' to bring them the news that 'Bob's gone'.[68] Writing many years later, William's daughter Gwen said of her mother that:

> She was always happy painting; nothing in life gave her so much direct pleasure . . . But the First World War and the loss of her younger son dealt her a grievous blow from which she never really recovered. Her gaiety went; her cheerfulness from then on was a fire lit for others, at which she warmed her own hands.[69]

Bob's death also brought a profound change to William's outlook. Before the war, he had felt that science transcended national borders and rivalries, but in the aftermath of Bob's death this optimistic spirit of internationalism drained away. When the Royal Swedish Academy of Sciences finally awarded the prizes for the years 1914–1919 at a ceremony held in Stockholm in 1920, having been unable to do so during the years of the War, neither William nor Lawrence was present to receive their honour. Publicly, William cited teaching commitments and 'several other engagements' as an excuse; but his casual remark, 'I believe that several Germans are going', made privately in a letter to Ernest Rutherford may well reveal his real reasons for not travelling to Stockholm to accept the award.[70]

The War had taken its toll emotionally on William, but it also marked an important change in the direction of his science and one that would lay the foundations for the work of his protégé Astbury. Since the end of the nineteenth century, there had been a growing feeling that British industry had become complacent and would easily be left standing by rival nations, the biggest of which was Germany. In 1910, an article in *Nature* pondered the question of how Prussia had been transformed into a successful industrial economy given that, only a hundred years earlier, it had been an impoverished state with little infrastructure, following defeat by Napoleon. The article concluded that there were two factors in this success story—education and cooperation.[71]

In the field of education, Germany boasted a vastly superior system of technical training for workers in manufacturing industries. *Nature* declared that:

> There is not a student of national economy who fails to realise that Germany and the United States, now serious rivals to English trade, owe their rapid industrial and commercial development largely to the magnificent system of technical education which they have established.[72]

An earlier article, reporting on a visit to German technical schools by a delegation from Manchester in 1897, noted that:

> Nothing struck the English visitor to Germany more than the extraordinarily large number of well-educated young men in the day departments of foreign technical schools, clearly pointing to the recognition of the value of scientific training as the chief element and necessity for industrial efficiency and success.[73]

The same report went on to warn that:

> It was quite time that the veil was torn from the eyes of the English workman, and that we abandoned the short-sighted belief that no-one could touch us in our various industries.[74]

The second factor, that of cooperation, was evident in the way Germany had recognised the need to support technical education by funding and encouragement from central government. Hailing the superiority of German organisation in the support of scientific training, *Nature* lamented:

> Instead of this, what do we find in England? The British Government has chosen the easier course of leaving the founding and management of technical institutions to the enterprise of charitable private persons, corporate bodies, and the local authorities.[75]

The result of this situation, it went on to warn, was a complete absence of organisation on the scale seen in Germany and resulting 'disastrous competition between the existing schools'.[76]

With the outbreak of war in 1914, yet more weaknesses in British industry had been exposed. The first of these was that British manufacturing was highly dependent on imports of raw materials from abroad.[77] The second was that, in addition to its superior organisation of technical education, not only had Germany recognised the need to harness basic scientific research with the needs of industry, but her financiers and industrialists were often trained scientists. This stood in sharp contrast to the situation in Great Britain, which *Nature* described in 1915 as the 'neglect of science in industry and in public affairs, which is characteristic of this country'.[78]

By the end of the War, *Nature* was offering the following advice: 'It is always wise to learn from your enemies when you can, and Germany has much to teach us concerning the manner in which Science may be made subservient to War and to the conditions which war produces'.[79] It was a piece of advice upon which both the British Government and scientists had already begun to act. In 1915, the British Government had announced the 'establishment of a permanent official organisation for the promotion of scientific and industrial research'.[80] Called the Department of Scientific and Industrial Research, this body was endowed with £1 million and its aim was to make British industry less dependent on the import of foreign raw materials. There was a growing recognition that 'our industrial ascendancy had been challenged, if not wrested from us, by the capacity for organisation displayed by our commercial rivals, particularly Germany'.[81,82] It was also hoped that this organisation would solve the problem of what was described as 'the essential individualism of the average English industrialist' who had 'no interest in research which did not produce tangible results within a year'.[83]

Recognition that basic scientific research needed to be harnessed to industry with the support of the government was also a lesson that William Bragg had already recognised at the start of the War. In a letter to his colleague Arthur Smithells, Professor of Chemistry at Leeds, he wrote:

> After the war, if all goes right, we know, as you say implicitly, that there is another struggle coming on, in which the organisation and efficiency and well-being of England must be considered from the point of view of what science can do to improve them ... The physical laboratory at Leeds should develop itself into a research laboratory for the industries of the Riding. Its professor and staff should have before them, as the ideal behind their work, the investigation of all physical questions that are involved in the textile trades to begin with ... You will not get your manufacturers here to approach it the right way without an example, they are too ignorant, stupid and jealous, with honorable exceptions.

I quite understand, as you know I do, the tale of qualities on the other side which have made Yorkshire the great industrial centre that it is.[84]

Of all the 'industries of the Riding' to which Bragg referred, it was the manufacture of textiles that he felt might benefit in particular from the application of physics. Heavily dependent on the import of dyes manufactured in Germany, the War had exposed the vulnerability of the textile industry, and a government report in 1918 identified textile manufacture as one area that would benefit significantly from more investment in basic scientific research.[85] But several years earlier, just after the start of the War, Bragg had already begun to consider what valuable new scientific insights a physicist might bring to the local textile industries of Leeds.

In 1923, *Nature* published the synopsis of a lecture delivered to the Institute of Physics entitled 'The Physicist in the Textile Industries'. It hailed the textile industry as 'one of the greatest factors in civilisation' but complained that 'it is not using to the full the immense powers bestowed on this generation by scientific discovery'. The textile industries, it proclaimed, 'offer an almost entirely unexplored and unlimited field for the research physicist'.[86] Eight years earlier, William Bragg had already drawn exactly the same conclusions and was convinced that the textile industries of Leeds would benefit greatly from the input of physicists such as himself, saying, 'the Textile department does not know enough physics: it could not be expected to. But the physics problems are not to be solved except by good men, they must be taken very seriously.[87]

As a product of his time, perhaps we can forgive Bragg for not having the vision to see that the task of applying new methods in physics to the study of textiles might be accomplished equally well by 'good women' as by 'good men', but what he did nevertheless recognise was that this goal might be better achieved if he were to leave Leeds. By 1915, he had become convinced that if Britain were to prosper in the future, then it must follow the German example by forging closer links between science, industry, and government. This would require a much closer relationship between scientists and politicians seeking their advice when making decisions over policy. Having won the Nobel Prize, William was now one of the most eminent scientists in the country, whose advice would be highly sought after, and he wanted to ensure that he was well placed to be actively involved in shaping the scientific future of Britain. Writing to Arthur Smithells, he outlined his thoughts:

> part of the same work can be done in London: and I think that is where I might hope to come in. The Government will want help: I believe they are sure to do a lot, and in time I might get a footing in their councils. Moreover there is the Royal [*Society*]. Could we not make it an advisory body, with more worthy work than that of reading papers? . . . I think I could help all the more because I have been here, and lived in the centre of these industries, and got to know something of the character of the men & the masters, their wonderful good qualities and their ignorance of certain aspects of their work. It seems to me all one thing, and that I may be serving you and the general cause better by going now.[88]

On 1 September 1915, William took up his new post as Quain Professor of Physics at University College London. Writing in the pages of the Leeds University magazine, *The Gryphon*, Smithells declared: 'Let us be quite clear about this; Professor Bragg has not left us for the honour of going to London . . . [he] has left us because he thinks that in London he can do better work.'[89]

Despite his enthusiasm to support the local industries of Leeds with basic science, William felt that his move to London would be better for the nation's scientific future in the long term. Nevertheless, before his departure, he offered the following advice to those at Leeds who were considering how to follow up his ideas of bringing physics to bear on the problems of textile manufacture:

> now is the time for action; and if we could find a keen young man to take my place and to throw himself into the work which I have outlined, that is I think the next stage. Let him prove to them [*the textile manufacturers*] against their own convictions that scientific examinations of their problems is the right thing: and chuck important solutions at their heads.[90]

After announcing the departure of William at a University of Leeds Senate meeting in June, the Vice-Chancellor had declared that 'no heavier loss could have fallen on the University'.[91] Yet William's departure for London would ultimately prove to be Leeds' gain, for it was in London that he found the 'keen young man' who would eventually continue the work that he had begun in Leeds. His name was William T. Astbury.

3

'A Keen Young Man'

After being discharged from the Army in 1919, Astbury returned to Cambridge and completed his studies, graduating with first-class honours. By this time, he had already developed a fascination for one particular area of physical science where the boundary between physics and chemistry became blurred. This was the study of crystal structure, and for a young scientist hoping to establish themselves in this field there was only one place in which to serve an apprenticeship—University College London (UCL) with Sir William Bragg as mentor.

Despite officially taking up his post at UCL in 1915 after his departure from Leeds, William Bragg had spent the rest of the war working for the Admiralty on methods of detecting U-boats. After the war, he resumed his post at UCL, but his tenure there proved to be short-lived. On his arrival, he found himself once again having to build up a research project as he had done at Adelaide and Leeds and was increasingly frustrated by financial complications, institutional politics, and bureaucracy.[1] As a result, when the position of Director of the Royal Institution (RI) became vacant in 1923, he quickly took up the post and Astbury came with him.

Founded in 1799, the RI had once been the scientific home of scientists such as Michael Faraday and Sir Humphrey Davy, but when Bragg's team arrived there in 1923 the prestige of these years was only a distant memory. Lawrence Bragg had now taken up a post at Manchester, but on visits to the RI, he recalled the vivid feeling of malaise that permeated the place, saying: 'it had the forlorn feeling of a harbour when the tide is out'.[2] The libraries were empty and unused, except for a lone man who arrived at 10 every morning and proceeded to sit all day long in the same armchair, while, according to Lawrence Bragg, the only indication of active scientific research was a vague rumour of someone trying to extract gold from seaweed.

The arrival of William Bragg's team brought a new lease of life to the RI. As the field of X-ray crystallography was still in its infancy, it offered a vista of subjects for study that was sufficiently broad for each researcher in Bragg's group to carve out their own particular niche without any sense of being in direct competition with their peers and thus avoiding the ferocious and often brutal rivalries that have come to define certain fields of scientific research today. Discussions, whether conducted in formal research seminars or in the tea room, were free and not inhibited by reservations born of career politics. Light relief came in the form of lunchtime table tennis games into which Astbury introduced an extra competitive edge by announcing that additional points would be awarded to anyone who could knock a matchbox off the table.[3] Much of this collegial spirit was fostered by Bragg's style of leadership. Shunning what would today be called 'micromanagement', Bragg preferred a more 'hands-off' approach, opting instead to simply pop his head around the door of the laboratory every so often to ask a member of staff a technical question relating to a lecture that he was preparing. It was a style of leadership that won him the respect and loyalty of his staff, for, as Astbury said:

> he was not an active participant in everything, naturally, or the origina-tor of every new idea, but he was the soul of everything. We who were his disciples and are his apostles had always the impulse to 'tell Bragg about it' wherever we were working, for we knew that we could talk to him, that he would listen, and that he would be just as thrilled as we were . . . There was no question of his making regular tours to discuss progress; but he might pop into your room any time to tell you or ask you something.[4]

Thanks to this intellectual environment cultivated by Bragg, many of the young scientists who had joined his team later went on to become renowned pioneers themselves in various aspects of X-ray crys-tallography. One of these was Kathleen Yardley, with whom Astbury collaborated on his first piece of scientific work in Bragg's laboratory in a partnership that proved to be highly successful.

Perhaps one reason that Yardley and Astbury formed such a success-ful partnership was that they shared a certain empathy, having both trodden a similar path from humble origins to Bragg's laboratory via hard work, determination, and scholarships. Yardley had been born in Newbridge, Southern Ireland in 1903 and was the youngest of ten chil-dren, four of whom had died in infancy. Having grown up in severe poverty, her mother brought the family to England in 1908 in the hope of providing a better future and also to escape the political instability of Ireland at that time.

Settling in Seven Kings, Essex, Yardley won a County Minor Scholarship to the County High School for Girls in Ilford, where she excelled academically. She later said that her intellectual curiosity and love of learning had been inherited from her father, a postmaster whose knowledge encompassed the antiquities of Peru and the birds of Western Australia thanks to his eager reading of books purchased from junk stalls.[5]

In addition to having inherited her father's love of books, Yardley also displayed an impressive ability to memorise facts—a skill that she developed by making herself memorise sermons from church services she had attended and then writing them out by hand.

This skill served her well during her final two years at school, when, having developed a particular interest in the physical sciences, she took classes in physics, chemistry, and higher mathematics at the County High School for Boys, as these classes were not offered at her all-girls school. Thanks to her formidable discipline and determination, she won a County Major Scholarship, having achieved distinctions in English, French, history, geography, botany, and mathematics, as well as being awarded the Royal Geographical Society's medal for the highest marks in the geography and physical geography papers. As a result of this impressive examination performance, Essex Education Authority offered to increase her County Scholarship if she stayed on at school in order to apply to Cambridge University. But Yardley was keen to go to university as soon as possible, and so, aged only 16, she went to Bedford College for Women in London, where she began studying mathematics. But her real passion was physics and, against all advice (particularly that of her old headmistress, who warned that she would have little chance of distinguishing herself in physics), she changed to physics at the end of her first year. Her decision to switch courses proved to be a wise one, and the fears of her old headmistress most certainly proved to be unfounded. Following her final examinations in 1922, she came top of the list of all students taking physics at University of London colleges— an achievement that did not go unnoticed by one examiner in particular.

Having marked Yardley's examination papers and being left highly impressed by her performance, William Bragg contacted her and offered her a place in his research team at UCL. It was to be the start of a distinguished career that saw Yardley, under her married name of Lonsdale, becoming the first female Fellow of the Royal Society in 1945, and also taking the post of Professor of Chemistry and Head of the Department of Crystallography at UCL in 1949. Nor was science the only field in which she distinguished herself. With vivid memories of sitting doing her homework by candlelight under a table for protection during Zeppelin raids on London during the First World War, Yardley was a pacifist from an early age.[6] Having joined the Quaker movement during a brief time in Leeds where her husband Thomas worked, she was devoted to the cause of international peace. After the Second World War, she made visits to the Soviet Union in the cause of promoting better understanding and to draw attention to the plight of political prisoners there. She also did charity work for prisons in the UK, a vocation that was born from her own experience as an inmate in Holloway Prison during the Second World War, where she served a brief term on account of having failed to register for civil defence duties.[7]

It is a testament to her strength of character that she was able to draw positive lessons from what must have been a very intimidating environment, but the experience left her with a talent for being able to talk to anyone. As her husband Thomas later said: 'Before prison it might have bothered her to go to Buckingham Palace. Afterwards, Holloway or Buckingham Palace were all the same.'[8]

One person with whom she had no difficulty in talking was Astbury. Yardley had fond memories of their time working together in Bragg's laboratory and described Astbury as the most colourful personality in the team, recalling how everything from discussions of crystal symmetry to politics or even just the lunchtime games of table tennis that were an institution in Bragg's group were all enlivened by Astbury's presence. Writing an obituary of Astbury many years later, she said:

> He took me under his wing and helped me in every possible way. I think it was partly his natural kindness and partly the fact that his intense happiness in his recent marriage to a lovely young Irish girl made him feel so fatherly towards all young people. He was so full of enthusiasm for his work that none of us could feel it to be drudgery, even though it entailed sitting for hours on end with one eye glued to a microscope taking readings of the movements of a gold-leaf . . . it was impossible not to rejoice with him; and he did achieve miracles with what would have seemed to most people intractable problems.[9]

One of these seemingly intractable problems was the question of how to describe and classify what crystallographers called 'space groups'. In the same way that a piece of wallpaper can be thought of as being generated by repetition, or, to use the more mathematically accurate term, 'tessellation', of a single basic pattern in two dimensions, a crystal can be thought of as being the repetition of a basic fundamental pattern in three dimensions. This most basic arrangement of atoms or molecules that, when repeated, generates the entire crystal structure is known as the 'unit cell' and, rather like the most basic shape in a wallpaper pattern, it can have a number of different possible symmetries. In other words, there are a fixed number of ways in which any particular unit cell can be rotated, folded onto itself, or moved in such a way that it appears to be unchanged. Depending on its precise symmetry, the unit cell of any given crystal can be classified as belonging to a particular symmetry or 'space' group.

Mathematicians had shown that any shape that can tessellate in two dimensions, like a wallpaper pattern, must belong to one of 17 possible symmetry groups based around translation and rotation of the shape. Now the question that crystallographers wanted to answer was: how many symmetry, or space groups, are there in three dimensions? The answer, thanks to rigorous mathematical work by Astbury and Yardley, was that all crystals must be made of a repeating unit cell that can belongs to any one of 230 possible space groups. They then showed how the particular X-ray diffraction pattern made by a molecule could be used to determine to which of these possible space groups it belonged.

William Bragg's interest by this time had moved on from the theory of X-ray diffraction to questions concerning the structure of organic compounds (i.e., those whose chemistry was based on carbon) and, as a result, he had to be persuaded with some difficulty by Astbury and Yardley to submit their work for publication. When their paper, 'Tabulated X-ray data for the examination of the 230 space-groups by homogeneous X-ray analysis', was finally published in the *Philosophical Transactions of the Royal Society* in 1924, it quickly became apparent that the work was of major practical benefit to anyone starting out in X-ray crystallography. Later published as the *International Tables for Crystallography*, Astbury and Yardley's work made understanding the geometry of crystals far more straightforward, and the tables are still used to this day.

The publication of the tables was a significant advance in the theory of crystal structure and a major first step for both Astbury and

Yardley in their respective careers. But Astbury's work at this time was not confined to the purely theoretical, and it was his experimental work at the RI that was to have even greater significance, for it marked the beginning of an idea that would come to define his entire scientific career.

These first experiments involved the use of X-ray diffraction to explore the structure of organic crystals. On their return to scientific research after the First World War, William and Lawrence Bragg had agreed to divide the territory of research in X-ray crystallography between them. Lawrence, now at Manchester, concentrated on X-ray analysis of inorganic crystals, while his father was far more interested in exploring how organic compounds formed crystals. Carbon compounds held a particular fascination for scientists as they were known to be the raw materials and end-products of metabolism in living organisms. The hope was that somewhere within the chemistry of carbon might lie some peculiarity that distinguished living and non-living matter.

One scientist who had been particularly interested in this question was Louis Pasteur. Although more famously remembered for refuting the idea that micro-organisms arose by spontaneous generation and developing the process of sterilisation named after him, the early days of his career were spent studying tartaric acid, an organic acid that occurs naturally in grapes and which often forms crystalline deposits at the bottom of a wine glass. What intrigued Pasteur was that the acid existed in two distinct forms that, although chemically identical, could be distinguished by their different effects on polarised light.

That light can be polarised is a commonly experienced effect, as it is the underlying principle by which sunglasses reduce the glare of the sun. An unpolarised beam of light can be imagined as being a collection of electromagnetic fields, the strengths of which oscillate in various different planes with respect to the direction in which the light is travelling. If the beam of light now passes through a filter that blocks out the oscillations in every plane except one, the resulting beam that emerges from the filter is said to be polarised.

Pasteur was intrigued by the observation that certain forms of tartaric acid could rotate the plane of polarisation in a beam of light. In other words, when a beam of polarised light emerged having passed through a solution of tartaric acid crystals dissolved in water, its plane of polarisation had been rotated by a fixed angle. Yet other forms, such as paratartaric acid, by contrast, which had exactly the same chemical formula, showed no such effect. Why this might be the case became apparent when Pasteur studied crystals of both forms of the acid under the microscope. Paratartaric acid was composed of tiny crystals that, while identical in shape, were exact mirror images of each other. On separating these two kinds of crystal, Pasteur showed that each was able to rotate plane-polarised light by a fixed angle in exactly the opposite direction to the other. This now gave a simple explanation of why paratartaric and tartaric acid differed in their effects on polarised light. While tartaric acid was composed of only one kind of crystal, paratartaric acid contained equal numbers of both mirror images. With one half of the crystals rotating plane-polarised light in one direction, and the other half in the opposite direction, their net effect was to cancel each other out, resulting in the absence of overall rotation.

Pasteur's studies of the optical properties of tartaric acid were the beginning of an important idea in chemistry—one that would prove to be the foundation of Astbury's work: that the physical properties of a substance are dictated not only by the particular atoms from which it is formed, but also by the three-dimensional arrangement of those atoms in space. The two different crystal forms of paratartaric acid that Pasteur had observed shared the same chemical formula, yet they rotated plane-polarised light in exactly opposite directions. In what way then did they differ to give rise to such a pronounced effect? In 1874, two chemists, called Jacobus Henricus Le Bel and Joseph Achille van't Hoff, independently suggested that, while two molecules of paratartaric acid might share the same chemical formula, the way in which their respective component atoms were arranged in three dimensions might differ. A helpful analogy is to consider a pair of gloves. While both the left- and right-handed gloves each have four fingers and a thumb, the fingers and thumb are laid out differently in space to make the gloves mirror images of each other that cannot be superimposed.

What Le Bel and van't Hoff were proposing was that, although molecules might be composed of the same atoms, these atoms could, like the fingers and thumb in the glove analogy, be arranged in 'left-' or 'right-' handed configurations, or, to use the correct terminology, 'l' and 'd'. These differing configurations of the same molecule are called *stereoisomers*, and, curiously, for many complex organic compounds, such as those occurring in living systems, only one form of each possible stereoisomer tends to be found. The small amino acid molecules that make up proteins in living systems, for example, are all of the l-form, while carbohydrates such as glucose in starch are all of the d-form. Quite why molecules within living cells should show this strong bias towards one particular stereoisomeric form remains a puzzle to this day. Yet stereoisomerism is far more than just an enigma for theoretical chemists, for it can have profound effects. One particularly tragic example of how stereoisomers can differ in their effects is the drug thalidomide. Developed in Germany in 1957, thalidomide was prescribed to thousands of women worldwide as a mild sedative. What was not realised at the time was that the thalidomide being administered actually consisted of a mixture of two different stereoisomeric forms, or molecules that were mirror images of each other. While one of these forms had the beneficial effect of countering morning sickness, its mirror image was a potent teratogen that gave rise to birth defects.

For Astbury, X-ray diffraction was the ideal tool to provide the experimental evidence that certain physical properties of a molecule might arise from differences in the spatial configuration of its atoms, or stereoisomerism. In particular, he hoped to show that the stereoisomers of tartaric acid had different effects on polarised light because the four carbon atoms of each molecule were arranged in a spiral shape that ran in opposite directions in each of the two different forms of the molecule.[10] It was an ambitious aim, and one that proved to be unsuccessful. Despite his best efforts using Bragg's X-ray spectrometer, Astbury was unable to show how the arrangements of atoms in the two forms of tartaric acid resulted in differing optical activity.

But, despite having failed in his first major piece of experimental research, this work was nevertheless highly significant for Astbury. John Desmond Bernal, another member of Bragg's young team who, like Astbury and Yardley, would go on to become a pioneer in the field, later noted that this piece of work marked a key moment in Astbury's development as a scientist. According to Bernal, this early paper was valuable not so much for its scientific contribution but for what it revealed about Astbury's thinking.[11] Within this paper was an idea that, when eventually applied to biology, would ultimately be the cornerstone of all his success as a scientist and come to dominate the rest of his scientific career: that physical properties of biological substances could be explained in terms of differences in the three-dimensional shape of their component molecules and the way in which these shapes changed; as Astbury himself later put it, 'It is the actual shapes of molecules that count in the end.'[12]

That Bernal expressed this insight many years later in an obituary to Astbury is testament to the close professional relationship that the two men shared, and it was one that had begun during these early days in Bragg's laboratory. Unlike Yardley, Bernal's origins contrasted starkly with those of Astbury. Three years younger than Astbury, he was the son of Anglo-Irish gentry from Brookwatson, Tipperary, and had won a scholarship to Emmanuel College, Cambridge, where he thrived. Bernal was a genuine polymath—widely read in arts, literature, history, philosophy, and mathematics, to the extent that he acquired the nickname that would last all his life: 'Sage'.[13] Of all these interests, though, two in particular truly fired both his intellect and his passion. These were science and Marxist politics, and it was while at Cambridge that Bernal embraced both. The materialist philosophy at the core of Marxist thinking held a particular appeal for Bernal, as it resonated with his personal view of science. The promise of Marxism to expose religion as nothing more than a delusion conjured up to bolster a particular economic status quo resonated with Bernal's view that science could offer an understanding of life purely in terms of molecular mechanism without any need to invoke a non-material or spiritual element. In forsaking the Catholicism of his childhood to take up this new stance, Bernal made an interesting contrast with both Yardley and William Bragg, neither of whom had any difficulty reconciling their scientific work with a private faith or at least a sympathy towards certain strands of religious thought. Unlike Bernal, William Bragg believed that there was no fundamental contradiction between being a scientist and holding religious beliefs and felt that the two could work in harmony. In answer to the charge that science and religion must be diametrically opposed, Bragg's response was to admit that they were indeed opposed, but only in the same way 'that the thumb and fingers of my hand are opposed to one another. It is an opposition by means of which anything can be grasped.'[14]

Bernal's Marxist convictions were not solely rooted in abstract philosophical ideas, however. Following the economic crash of 1929 and the subsequent rise of fascism during the 1930s, he felt that they provided a political solution to world events. While he himself played an important role in fighting fascism by providing scientific advice in planning the D-Day landings of 1944, Bernal always felt that the real solution

to the problems of the world was already to hand. For him, the Soviet Union was a shining example of how, once science was harnessed
by the State, human society could be organised according to rational,
planned directives. Unlike many other radicals of the Far Left, Bernal's
enthusiasm for the Marxist–Leninist cause never abated with age. After
the Second World War, he made regular visits to communist countries
and was a vocal advocate of nuclear disarmament. Even after the brutal Soviet suppression of the Hungarian uprising in 1956 caused many
on the Left to view the USSR with slightly more suspicion, Bernal
remained an enthusiastic supporter.[15] Nor were his enduring and undiminished passions confined only to the political Far Left. An obituary
once described him as having 'lived life to the full', which was probably a tactful and polite reference to a number of alleged extramarital
dalliances.[16,17]

When Bernal joined Bragg's team at the RI in 1923, his first attempts
at experimental work were spectacularly unsuccessful, with the only
results being burns to himself and broken equipment. Kathleen Yardley
diligently instructed him in the use of the X-ray spectrometer, but he
soon gave up, feeling that he lacked the necessary patience. After plead-
ing with Bragg to be allowed to work on theoretical problems, he was
assigned the task of using X-ray diffraction to probe the atomic structure of graphite. As this work began to yield results, Bernal felt that
he had become an active member of the laboratory and described how
Bragg's team used to hold weekly colloquia that were held in Bragg's
flat on the top floor of the RI. These meetings were usually dominated
by Astbury, who, according to Bernal:

> was always brimful of ideas but often these were rather difficult to under-
> stand. When he spoke, most people thought he was talking nonsense.
> I found out fairly early that when Astbury was talking, it might ap-
> pear nonsense, but it always contained a valuable and new idea and I
> did my best at these meetings to interpret them and, what was much
> more difficult, to get Astbury's agreement that I had interpreted him
> correctly.[18]

One of these new ideas was to have a profound effect on Astbury's research and also that of Bernal. In all likelihood, Astbury could easily
have spent the rest of his career working on X-ray analyses of simple organic compounds such as tartaric acid, and in so doing achieve
a modest degree of success, had not Bragg casually asked one day, as

was his habit, whether Astbury could help him prepare some X-ray photographs for his forthcoming 1926 Royal Institution Lecture. The lecture was to be called 'The Imperfect Crystallisation of Common Things'.[19] The 'common things' in which Bragg had become interested were not simple crystals like tartaric acid, but the naturally occurring fibrous structures like hair that were found in living organisms. By the time of Bragg's lecture in 1926, work on the structure of natural fibres had already been underway in Germany. Before the First World War, Kaiser Wilhelm II had declared that a mighty nation like Germany, renowned for its scientific prowess, should have its own centres of scientific excellence to match the prestige of places like the RI in Great Britain, the Pasteur Institute in France, and the Rockefeller Institute in the US.[20] As a result, a series of scientific research institutes were established and named in his honour, with the goal of promoting pure science. It was at one such institute, the Kaiser Wilhelm Institut (KWI) for Faserstoffchemie ('chemistry of fibres') established in 1920 at Dahlem, a suburb of Berlin, that the first X-ray diffraction patterns of naturally occurring fibres such as rubber and cellulose were obtained. From the resulting diffraction patterns, it became evident that, while these fibrous substances did not contain regular repeating structures in three dimensions (as in a crystal like sodium chloride), they did contain regular repeating regions of structure along one dimension. These regions, called crystallites, were aligned along the axis of the fibre and had enough definition and order to give a meaningful X-ray diffraction pattern from which important information about the molecular structure of the fibre could be deduced.

The implications of this work were enormous. In these ordered patterns of spots lay the tantalising possibility that X-rays might be able to probe not only the atomic structure of simple organic and inorganic crystals such as tartaric acid and sodium chloride, but also the molecular architecture of the complex biological structures found within living systems. William Bragg felt that the study of biological structures would be the crowning achievement of X-ray crystallography, and, in preparation for his lecture, he assigned Astbury the task of taking X-ray photographs of a diverse range of natural fibres, including human hair, cotton, and spines from sea urchins. Despite later describing the quality of his pictures as 'pretty dreadful',[21] Astbury quickly recognised that within the resulting diffraction patterns lay important clues to the hidden molecular structure of these materials.

Astbury now surely felt the same exhilaration that Lawrence Bragg had had described when he and his father had performed the first studies on simple crystals using the X-ray spectrometer. Here before him lay a whole new vista of research possibilities. That natural fibres gave regular diffraction patterns suggested that not only could non-living matter be explored using X-rays, but also the molecular architecture of life itself might be amenable to analysis by this method. The immediate question was where to continue this work, for Bragg now felt that the time had come for Astbury to step out of his shadow and establish himself as an independent scientist.

An ideal opportunity presented itself in 1927 when the new post of Lecturer in Structural Crystallography was created at Cambridge. Astbury applied but was rejected, possibly because, despite having taken the photographs for Bragg's RI lecture, he had not yet published any of this work in a journal.[22] Perhaps it was also due to his somewhat brusque manner during the interview. When at one point he was asked how he felt about scientific collaborations, he replied curtly: 'I am not prepared to be anybody's lackey'.[23]

Bernal applied for the same post and fared somewhat better. Throughout the interview, he responded to all questions with curt 'yes' or 'no' answers, but when asked what he would do with a department of crystallography if offered the post, 'he threw his head back, hair streaming like an oriflamme, began with the word No . . . and gave an address, eloquent, passionate, masterly, prophetic, which lasted for forty-five minutes'.[24] As Professor Arthur Hutchinson, who was head of the Department of Mineralogy and sat on the interview panel later remarked, 'There was nothing for it but to elect him.'[25]

At the end of the first academic year in his new post, Bernal made a tour of several laboratories in Europe that were doing crystallographic work.[26] While on this trip, he visited the laboratories of the company I. G. Farben at Ludwigshafen in Germany, where the scientists who had originally done the X-ray work on natural fibres at the KWI were now based. Here he learned much more about work done by the Austrian scientist Hermann Mark and his Swiss colleague Kurt Meyer, who had used X-ray diffraction studies of cellulose fibres to build a theoretical model of what the molecule might look like. According to his biographer Andrew Brown, Bernal was 'entranced' by Mark and Meyer's work, such that on his return to Cambridge he began to ponder how he might himself apply X-ray diffraction to the study of biological molecules.[27]

Yet while Bernal returned to England inspired by the studies of Mark and Meyer, it was Astbury who would ultimately reap the most reward from their work. Of all the various natural fibres studied in Germany, two were of particular significance to Britain: cotton and silk. These were both raw materials of the German textile industry, and this work was yet another example of how Germany excelled at harnessing basic science with the needs of the economy—a talent that, as mentioned, was causing much concern in Britain at this time. The Department of Scientific and Industrial Research, which had been formed by the British government during the First World War in response to these growing concerns, had established 13 associations, the aim of which was to support basic scientific research in a particular industry.[28] One of these was the British Research Association for the Woollen and Worsted Industries, a name later simplified in 1930 to the Wool Industries Research Association, or WIRA.[29] Between 1917 and 1919, this association actively supported research in the Department of Textiles at the University of Leeds, which was the largest department in the university and one of the biggest in the country.[30]

But in 1920, WIRA acquired a four-acre site on Headingley Lane, which lies between the University of Leeds and the world-famous Test cricket ground, close to William Bragg's first residence. Here, a grand old Victorian house called 'Torridon' was converted into a research institute. The drawing room became a chemistry laboratory, while physics was to be done in the billiard room and the wine cellar used for work involving controlled temperature and humidity.[31] With these facilities at their disposal, WIRA could now conduct its own research independently of the University.

The result was that, while Leeds boasted the largest Department of Textiles in the country, the department was now chiefly concerned with teaching, not with research. This was a situation that was deemed to be unacceptable, particularly if Leeds wished to retain its reputation as being the national centre for work in textiles.

In 1927, J. B. Speakman, the only graduate in the department other than its professor, Aldred Barker, expressed his concerns for the department in a letter to the Vice-Chancellor of Leeds. According to Speakman, the main problem was that teaching in the department lacked any grounding in basic science, with the result that the students understood nothing of the scientific principles underlying their work. Only by more detailed knowledge of the chemical and physical properties of wool could both the education of the students and the productivity of the industry be improved. And Speakman believed that he had an idea of how this could be achieved. 'I was led to the view', he said, 'that X-ray analysis would be of service', and he went on to suggest that it would be through the appointment of a physicist that the problems of the Department of Textiles might be solved.[32]

In May 1928, Barker wrote to William Bragg asking whether he knew anyone who might be suitable for the new post of textile physicist at the University of Leeds that had been created in response to Speakman's suggestion with financial support from the Worshipful Company of Clothworkers.[33] In response, Bragg wrote back saying that he had the perfect person for the job.[34] In his letter, Bragg described Astbury as 'a brilliant man very energetic and persevering, has imagination, and, in fact, he has the research spirit'. Although he went on to say that he was reluctant to lose Astbury, Bragg must have been delighted. Ten years earlier, Speakman had written to the Vice-Chancellor of Leeds saying that the most important research into textile fibres had received its impetus thanks largely to the ideas of the recently departed Cavendish Professor of Physics, William Bragg.[35] In Astbury, Bragg had at last found the 'keen young man' who could return to Leeds and continue his legacy there.

4

'Into the Wilderness'

Unfortunately, Bragg's enthusiasm that Astbury should move north was not shared by his protégé. Astbury's work at the Royal Institution in London had held the promise of breaking exciting new ground in chemistry at a fundamental level. The work at Leeds, however, was to be very different. As the newly appointed lecturer in textile physics, Astbury's task would be to perform X-ray studies of wool fibres for the local textile industries and, as was evident from a letter written to J. D. Bernal at the time, this was not a prospect that thrilled him:

> I don't know whether you have heard the sad news, but it seems possible that I have abandoned crystallography. I have accepted the new lectureship in Textile Physics at Leeds University and am leaving the Davy-Faraday at the end of the present month. I am making one last despairing effort to keep in touch with crystallography by attempting to do some X-ray work on wool, etc. and I want to get some apparatus together.[1]

The thought that his best days of genuine scientific research might already be behind him just as he starting to establish himself as an independent researcher left him feeling as if he was 'going into the wilderness'.[2] To add to this frustration, on arriving at Leeds, he was shocked and rather irritated to discover that someone else there was already using X-ray methods to study wool.

Over the previous 18 months, J. Ewles, Lecturer in Physics at Leeds, had made some tentative first studies of wool fibres using X-ray diffraction methods, which he had presented when the British Association held its meeting in Leeds in 1927. But Ewles' photographs were of poor quality and blurred, leading him to dismiss them as being of no value. It was a judgement for which Astbury would be forever grateful. Seizing the opportunity that Ewles' inertia provided him, Astbury acted quickly. Feeling that the study of wool fibres using X-rays was his rightful territory and that it had been encroached upon, Astbury appealed to the chemist J. B. Speakman—one of the leading figures in the Department of Textiles—who was studying the chemical properties of wool and, following Astbury's appeal, decreed that all further work on the physics of wool should be carried out by Astbury alone.[3]

With this swift and amicable coup, Astbury established his domain. But it was one that he had now largely to build from scratch. With funds that had come mostly from the Worshipful Company of Clothworkers, he was joined by a mathematician H. J. Woods and two PhD students; he was given an allowance for expenditure on apparatus of £150 per annum. The first four months of his appointment were spent converting a bare room into something that could reasonably be called a research laboratory.[4] Working very much in what J. D. Bernal called 'the great sealing wax and string tradition', all the equipment, such as the discharge tubes for generating X-rays, had to be built by hand.[5] Thankfully, Astbury did not have to contend with the problems that Lawrence Bragg faced when trying to obtain reliable equipment for his first X-ray studies from the workshop at Cambridge. The mechanic at Leeds, Mr Binks, was highly skilled and able to construct the sophisticated equipment that Astbury's team needed to begin their work.

No amount of technical skill on the part of a workshop, however, could reduce the laborious and time-consuming process of taking an X-ray photograph, not to mention the discomfort and danger involved. The bare room that Astbury had converted into a laboratory had the unfortunate feature of having one wall that was entirely glass and, owing to the particular layout of this room, it was against this wall that the discharge tubes that generated the X-rays were sited. Since the whole procedure of taking an X-ray photograph required complete darkness, curtains had to be hung over the glass wall to block out all daylight that would have otherwise fogged the photographic plates. In order to draw the curtains, however, a worker had to squeeze past the exposed end of the discharge tubes. This was more than just an inconvenience—it was also quite dangerous as it meant coming into close contact with the X-ray tube, which could reach temperatures in excess of 2000 degrees Celsius and also contained electrical voltages as high as 150 kilovolts.[6]

In addition to the risk of serious burns or electrocution, there was also simply the sheer tedium involved in taking an X-ray photograph. The pressure of the gas in the discharge tube that generated the X-rays had to be neither too high nor too low, and evacuation to just the right pressure was very much achieved by trial and error. Furthermore, the discharge tubes had to be regularly checked for leaks that would cause the pressure to drop too low. Also, because of the high temperatures generated in the tube, the cathode suffered regular damage, which meant that it had to be either repolished or completely replaced on a daily basis. This could be a delicate and painstaking operation, but by far the biggest headache was that the focal point of the X-ray beam could vary between individual experiments. That is, before any actual experimental work could begin, the focal point of the beam had to be located and the camera set-up adjusted accordingly—a laborious process carried out in the dark that could take anything between a few minutes and several hours. Given that the exposure required for a decent X-ray photograph of a fibre specimen could be several hours in length, very little could be achieved in a working day unless the equipment was all up and running by 8.30 a.m. When other groups began to work in this same field, Astbury remarked: 'They'll have to get up very early in the morning if they want to beat us.'[7] It was less a comment on the scientific calibre of his team, and more a simple statement of fact about the painstaking nature of X-ray work.

Far from being a sophisticated and elaborate piece of equipment, the 'camera' with which Astbury carried out this work was constructed by hand and consisted of a box made of three lead walls to shield the workers from the scattered X-rays (Figure 10).[8] To begin the studies, a carefully prepared 1-mm-thick fibre of Merino or English Cotswold wool was mounted inside the box and the beam of X-rays directed onto the sample. The resulting scattering of the X-rays was then measured using a photographic film placed behind the sample. In addition, a range of other natural fibres, such as human hair, llama hair, hedgehog spines, and porcupine quills, were also tested. Although this might at first sight appear to be a rather eclectic range of materials, they all shared one feature in common: like wool, they were made of the protein keratin.

The test fibres were subjected to a range of varying physical conditions, including tensile stress, temperature, and humidity, and the resulting patterns made by the diffracted X-ray beams as they passed through the fibres were recorded onto photographic plates mounted behind the sample. Astbury and his assistant A. Street took over 100 photographs in this way—a rather laborious task considering that each single exposure might well take up to 10 hours.

(a)   (b)

(c)

Figure 10 Astbury's X-ray camera. At first sight little more than a lead box with one side missing, it was with this hand-made apparatus that Astbury made his first X-ray diffraction studies of the molecular structure of wool, and eventually nucleic acids.

Photographs by K. T. Hall.

Wool might well have been useful for knitting jumpers and scarves, but it seemed far less promising as a subject on which to build a prestigious and successful scientific career. As Astbury himself said, 'all we knew was that certain clothes were made of wool, and that wool in turn was composed of an "amphoteric colloid" called keratin—a biochemically lifeless and uninteresting protein'.[9] Yet, when Astbury published this work in a series of papers from 1930 to 1935, they caught the imagination of the international scientific community at once, for they provided an answer to the question of what proteins actually looked like at the molecular level—their structure and composition—something that had long left biochemists puzzled and perplexed.[10–15]

Proteins are very much the molecular Cinderella of modern biology. Thanks to the iconic status of the double helix in popular culture, it is well recognised outside the scientific world that DNA is central to biology as it carries hereditary information and the instructions to make new cells. But what is often less well appreciated is that proteins are equally vital in the operation of the cell. Although DNA may be the carrier of hereditary information, it is only thanks to myriad different protein molecules that this information is executed and processed in the cell.

Some proteins such as Astbury and Street's keratin fibres have a purely structural role—like molecular girders, they form the physical building blocks of cells and tissues. But many other proteins function as molecular machines carrying out operations essential to the functioning and maintenance of the cell. Enzymes, for example, act on specific target molecules called substrates to catalyse chemical reactions that would never normally occur in a test tube. One example is amylase, the digestive enzyme found in saliva, which is able to cleave precisely the chemical bonds between the glucose molecules that are joined together to make starch. Other proteins are involved in the transport of essential nutrients around the body. For example, haemoglobin found in red blood cells consists of four individual protein chains, each of which has at its centre an atom of iron that enables it to bind to oxygen in the lungs. Once bound, the oxygen is transported via the bloodstream to sites in the muscles and lungs, where it is released for the generation of energy. Yet another essential function carried out by proteins is to convert one form of energy into another. The protein rhodopsin, found in the rod and cone cells of the retina at the back of the eye, is able to trap photons of visible light and convert them into electrical energy, which is relayed via the optic nerve to the visual cortex of the brain. With such impressive molecular machinery, Nature has evolved nanotechnology long before we did.

By the end of the nineteenth century, chemists had shown that proteins were composed of smaller molecules called amino acids. Chemical analyses showed that different proteins were composed of up to as many as 20 different types of amino acid. But what remained a mystery was how these different amino acids were organised to form a protein. Towards the end of the nineteenth century, the German chemist Kekulé had proposed that certain organic compounds—for example, benzene—were actually cyclic rings formed by carbon atoms bonded to each other. Kekulé then went on to suggest that carbon atoms might also bond with each other to form giant long chains. In 1902, this idea was taken up by another chemist, Emil Fischer, who applied Kekulé's idea to biological compounds like proteins. Knowing that two amino acids could be chemically joined together in a linkage called a peptide bond, Fischer speculated that proteins might be long chains formed by individual amino acids joined together like beads on a string through the action of peptide bonds to form what he called a 'polypeptide'.

However, another school of thought pointed out that many of the physical properties of proteins, such as their viscosity, were typical of colloids—loose aggregates of small molecules. According to this view, proteins were very different to ordinary small molecules. While every molecule of methane always has one carbon atom bonded to four hydrogen atoms, or a molecule of ammonia has one nitrogen atom bonded to three hydrogen atoms, proteins possessed no such regularity. As a loose collection of amino acids, they could have no specific size or mass. The problem, however, was that, while proteins might exhibit the physical properties of colloids, it was difficult to explain how a loose aggregate of amino acids could repeatedly perform a specific and well-defined function such as the binding and transport of oxygen by haemoglobin.

Although the colloid theory of proteins was popular, not everyone was convinced. One such scientist was German chemist Hermann Staudinger, who would later go on to win a Nobel Prize. Taking his inspiration from Kekulé and Fischer, Staudinger proposed that, like methane or ammonia, each type of protein molecule did indeed have a regular defined molecular structure and that this structure was huge— so huge, in fact, that Staudinger coined the term 'macromolecule' to contrast a protein with a small molecule like methane or ammonia. His ideas were not well received. When he presented his work in a lecture to the Zurich Chemical Society in 1925, the meeting ended with him shouting the words of the Protestant reformer, Martin Luther, in defiance at his critics: 'Hier ich stehe, ich kann nicht anders' ('Here I stand, I cannot do otherwise').[16]

Evidence in favour of Staudinger's model of proteins as giant chain molecules of fixed length was gaining momentum, with the support coming largely from physics. In 1924, the Swedish chemist Theodore

Svedberg developed a method for measuring the molecular weight of large compounds such as proteins using ultracentrifugation. Svedberg had found that when protein mixtures were placed in a dense liquid medium and subjected to accelerations 100,000 times the strength of gravity, the proteins moved through the medium until they reached an equilibrium point dictated by their size. The larger the protein, the further it moved through the medium when spun at high speed. If proteins really were colloidal aggregates of variable length, then it might be expected that this would give rise to a poorly defined diffuse band. Instead, when Svedberg subjected haemoglobin to ultracentrifugation, he found that it gave rise to a sharp, tight, well-defined band, suggesting that the haemoglobin protein had a fixed molecular weight and was therefore of a specific size.[17]

At the same time as he was working on ultracentrifugation, Svedberg also attempted to develop another method of measuring the size of proteins based on the application of an electric field. The theory behind the method was that, when placed in an electric field, larger proteins would move more slowly than smaller ones and so would separate according to size. Called electrophoresis, this method was successfully developed by one of Svedberg's students, Arne Tiselius, who used it to show that different types of protein had a fixed size.[18]

Other evidence came from the study of enzymes—the catalysts that speed up metabolic reactions inside the cell that would never normally occur in a test tube. In 1926, Cornell University biochemist James Sumner obtained crystals of the enzyme urease and, four years later, his colleague John Northrop showed that the digestive enzyme pepsin was also a crystalline protein. In both cases, the fact that the enzymes could be crystallised was strong evidence that they must both be proteins with a highly defined and regular structure.[19]

Staudinger's defiant stance was further vindicated a couple of years later by the results of the Austrian chemist Hermann Mark, whose work with his colleague Kurt Meyer at I. G. Farben had been such an inspiration to Astbury's former colleague at the Royal Institution, J. D. Bernal. Mark and Meyer had used X-ray analysis to show that cellulose, one of the main fibres in the plant kingdom, was composed of long chains formed by repeating units of the sugar glucose. The basic structural unit that Meyer and Mark detected by X-ray analysis was formed of two glucose molecules chemically bonded together.[20–22] The two glucose molecules were folded up in two dimensions to form hexagonal rings and rotated through 180 degrees to be mirror images of each other. Through continuous repetition of this basic structural unit, the entire length of the cellulose chain was built up.

Meyer and Marks' model of cellulose was a gift to Astbury. The X-ray patterns that they had obtained from cellulose fibres looked remarkably similar to those that Astbury had obtained from keratin fibres. Astbury reasoned that if cellulose was a giant chain molecule made up of smaller repeating glucose units, then, by analogy, might not keratin also be a giant chain of repeating amino acids—rather like the way in which stringing individual beads together can form a necklace?

From the patterns of smeared, dark spots on their X-ray diffraction patterns, Astbury and Street deduced that the keratin protein of wool and hair must be made of regular structural units that repeated at a distance of every 5.15 angstroms (an angstrom is one ten-millionth of a millimetre) to form a giant molecular chain. Their blurry photographs provided experimental confirmation of the hypothesis first proposed by Fischer in 1902 that proteins were not colloidal blobs of amino acids in a loose aggregate, but a string of amino acids chemically bonded together in a specific way to form a polypeptide—or, as Astbury preferred to call it, a giant 'molecular centipede'.[23,24]

With their X-ray photographs of keratin, Astbury and his team had answered a fundamental question in biochemistry. Out of their five papers published at this time, one in particular had profound implications that went far beyond West Yorkshire's woollen mills: stretched and unstretched fibres of wool keratin gave two very different X-ray diffraction patterns. Later hailed by J. D. Bernal as 'the key paper of all Astbury's work', the paper not only confirmed what proteins *were*, but also gave the very first hint at how they might *work*.[25] From the X-ray pattern of unstretched keratin, Astbury and Street calculated that the amino acid chain of the protein was folded up into a compacted state, which they called the 'alpha form'. When the wool was stretched by about 90%, however, the X-ray patterns changed in a subtle but significant way, suggesting that the chain had changed shape. In a movement rather reminiscent of a child's slinky toy, the chain had extended from a compacted form into an elongated, stretched form, which came to be known as the 'beta form' (Figure 11).

When Astbury had first started using X-rays to probe the atomic structure of tartaric acid crystals at the Royal Institution, he had held high hopes that the physical properties of this compound could be

alpha - keratin   beta - keratin

Figure 11 Astbury's X-ray studies of keratin fibres in wool showed that they were composed of long-chain molecules that could assume either a more compact 'alpha' form, or an extended 'beta' form and that the reversible transition between the two gave a molecular explanation of why wool was elastic.

explained in terms of its molecular shape. In this, he had been unsuccessful. The molecular shape of tartaric acid crystals gave no explanation as to their effect on polarised light. However, keratin was different: Astbury's description of how the amino acid chains of keratin changed shape from the compacted alpha form to the elongated beta form gave an explanation in molecular terms of the one physical property of wool that made it of such interest as a textile fibre—its elasticity. Despite also being a long chain of amino acids, silk did *not* exhibit the same elasticity, because, as Astbury's X-ray photographs showed, the amino acid chains of silk were identical to those of beta-keratin from wool. The polypeptide chain of silk was therefore in a permanently extended form and did not have a more compacted form like alpha-keratin in wool. It was an insight that was expressed more wittily by the mathematician Lindo Patterson:

> Amino acids in chains,
>   are the cause or so the X-ray explains,
>   of the stretching of wool,
>   and its strength when you pull,
>   and show why it shrinks when it rains.[26]

With this work on keratin in wool, Astbury had given the very first account of how the function and properties of a biological material might be explained by a *change in the shape* of its chain molecules. This was a scientific milestone and, though novel at the time, it has come to lie at the heart of our understanding of how proteins work as biological nano-machines. The efficiency with which haemoglobin is able to bind oxygen in the lungs and then release it in the required tissues is all due to subtle changes in the shape of the four polypeptide chains that make up the haemoglobin protein; similarly, when rhodopsin in the eye absorbs a photon of light energy, this triggers a conformational change in its polypeptide chain that, in turn, is transmitted to adjacent proteins that regulate the flow of electrically charged ions across the cell membrane. The induced conformational change alters the flow of electrical charge across the membrane, which is relayed as a nerve impulse to the brain. As a final example, the ability of the enzyme lysozyme to kill bacteria by specifically cleaving the chemical bonds between units of sugar in the polysaccharide chains of their cell walls is because the polypeptide chain of lysozyme is folded up in such a way that its shape is complementary to the shape of the bacterial sugar residues. Like a key fitting into a lock, the respective shapes of the enzyme and its target molecule fit together perfectly. In all these examples, it is the shape and precise folding of protein chains that determine their functions, an idea that, according to his colleague and close friend R. D. Preston, can be traced directly back to Astbury's work on wool, for it was from this work said Preston that:

> the idea was born, permeating the whole of protein investigation ever since, that what is essentially one and the same molecule can exhibit different features and properties according to the way it is folded.[27]

For Astbury, this emphasis on the importance of the shape of giant molecular chains in understanding biology defined a whole new approach to the study of living systems. It was an approach that he popularised with the term 'molecular biology', and offered a definition of this new science in his 1950 Harvey Lecture, 'Adventures in Molecular Biology':

> It [*molecular biology*] is concerned particularly with the forms of biological molecules and with the evolution, exploitation and ramification of these forms in the ascent to higher levels of organization. Molecular biology is predominantly three-dimensional and structural.[28,29]

Rather than be concerned with classifying organisms and arguing about their position in a taxonomic ranking, these new 'molecular' biologists would be a different breed entirely. For them, biology was to be done using the experimental apparatus of physics and chemistry, and would thereby become a quantitative, mathematical science. In a letter to Bernal, Astbury gave what has been called the first job description for a molecular biologist:

> He must be a biologist, of course—perhaps a physiologist would be best, but he must not be the romantic biologist, but one who thinks, or wants to think, in terms of molecular structure, and is not likely to wreck my apparatus, and will not ask me too often what cos θ is.[30]

Structure and shapes might well be at the heart of the new science of molecular biology, but Astbury was nevertheless still very keen to know more about the individual amino acids that made up the keratin chain. It was, after all, quite possible that understanding the individual components of the chain might give new insights into the overall structure and shape of the chain itself. To tackle this question, he made some initial chemical analyses of the individual amino acid components of the keratin chain and hoped to obtain crystals of some of them for study by X-ray analysis. But in this, his former colleague J. D. Bernal was way ahead of him.

At Cambridge, Bernal had obtained the first beautiful X-ray diffraction photographs showing that two individual molecules of the amino acid glycine were chemically bonded together to form a flat planar hexagonal structure, which was yet more proof that amino acids could form long chains of protein. While undeniably pleased at this experimental support for his own work, Astbury was also slightly disappointed. Having triumphed with his X-ray photographs of the keratin chains, he had harboured hopes of publishing the first X-ray diffraction patterns of the individual amino acids that made up these protein chains as well—but Bernal had beaten him to it. In a letter to Bernal, he betrayed a mild sense of irritation:

> I have only one grumble to make, and that is that you have done the diketo-piperazine after all, and just as I was meditating an attack on it. I have had the crystals of this extremely fundamental substance quite a year now, and since my structure of wool is more or less based on an indefinite repetition of this ring, I had come in a way, to regard it as my

own private property, to be analysed simply and solely by myself. However, I suppose I must not be selfish, especially as the structure which you have got out is so thrilling. You undertook to examine all the amino acids with the exception of the diketo-piperazine, so I shan't forgive you now until you carry out the investigation to the bitter end.[31]

Astbury was forced to admit that Bernal's work in this area was superior to his own. Rather than try to compete with Bernal on this front, the two of them reached a 'gentleman's agreement' according to which Astbury would concentrate on the structure of fibrous proteins, while Bernal would work on the crystal structure of the individual amino acids that made up these fibres. Writing to Bernal, Astbury suggested that this compromise was in both their best interests, as it meant that, by joining forces, they could stave off competition from elsewhere:

I have already collected specimens of most of the important amino-acids found in hair proteins and, as I told you, have got some sort of structure out for cystine . . . But if I send these amino acids on to you, will you harden your ridiculously soft heart and stop doing odd jobs for other people? That would be an essential condition, of course, if we are to keep this job in our hands and out of the reach of those damned Germans (God bless 'em!) . . . You ask me if I am going to stick to the proteins of animal fibres, but of course in a job like this it is impossible to stick to anything. It is not crystallography, but a kind of higher detective work, searching for clues, however faint, day after day . . . I would like to try everything I can lay hands on, but I mustn't be greedy. I will write to you further about these things, for frankly feel considerably safer with you on my side, so long as you are not developing my photos![32]

At the same time, however, some of Astbury's correspondence with Bernal does betray a vague suspicion that Bernal had the potential to be a formidable rival. Astbury's insistence that the two of them demarcate very clear boundaries to their respective investigations into protein structure suggests that he feared Bernal could all too easily become a competitor who might overtake him and, therefore, needed to be kept on a tight leash:

It is a fine bit of work which you have done, and I feel that together, me at the proteins, and you at their constituents, we should be able to knock a considerable hole in the subject. But I insist that you stick to the constituents, and don't 'snaffle' any proteins . . . I am convinced that structures quite analogous to that of hair are the basis of all proteins.[33]

With Astbury working on the structure of fibrous proteins, and Bernal making crystallographic studies of individual amino acids, it seemed that between them they had the entire field of research into protein structure under their control. But there remained one huge unanswered question. Fibrous proteins were only half the story, for there was a second group of proteins whose structure remained a complete mystery.

These were the globular proteins, such as the enzyme lysozyme and the blood protein haemoglobin. Unlike the fibrous proteins, the globular proteins were soluble in water. The mystery was what shape their polypeptide chains took to allow them to do this. It was highly unlikely that they would be soluble if their chains were simply long and straight like those of a fibrous protein such as keratin. Rather, the chain of a globular protein must be folded up into some complex, elaborate conformation, and X-ray diffraction was the ideal method to show this. Astbury had said to Bernal that he saw no reason why he should limit his work only to one half of the protein family, and so, on obtaining a sample of pepsin, he tried to study its structure using X-ray diffraction. But he was puzzled to find that, despite being a soluble, globular protein, it gave an almost identical diffraction pattern to the extended form of the keratin chain.[34] It was a result that made no sense, as it suggested that pepsin was also an insoluble fibre like keratin; yet the fact that it was secreted as a soluble enzyme by glands in the human stomach suggested that quite evidently it was not.

Bernal, meanwhile, had also obtained a sample of pepsin and given it to his student, Dorothy Crowfoot, to study. Later in her career, under her married name Dorothy Hodgkin, she went on to become one of the major figures in X-ray crystallography and won the 1964 Nobel Prize in Chemistry for her studies of the structures of vitamin B_{12} and penicillin, before solving the structure of the hormone insulin five years later. And the roots of her success had begun with these initial studies of pepsin. For, like Astbury, Bernal and Crowfoot also found that the pepsin sample gave a diffraction pattern very similar to that obtained from the fibrous, insoluble beta form of keratin—but what they noticed was that this pattern was obtained only when the pepsin sample was dry. By contrast, when the sample contained a certain amount of water, it crystallised, giving rise to beautiful diffraction patterns.[35] Their conclusion was that pepsin, like keratin, seemed able to assume two distinct physical forms, but whereas for keratin both forms were fibrous, pepsin could assume either a soluble, globular form or an insoluble, fibrous form, depending on its water content.

Bernal's results with pepsin prompted Astbury to have another attempt at X-ray analysis of a globular protein. But this time he chose a different subject. On one of his regular visits to Imperial College, London, he asked his collaborator there, A. C. Chibnall, whether it would be possible to have a sample of the globular seed protein edestin. Chibnall assigned his research fellow Kenneth Bailey, who had recently joined Imperial, the task of trying to obtain crystals of edestin for Astbury. Though initially unsuccessful, this was to be the start of a highly successful working partnership that resulted in Bailey eventually joining Astbury's team despite remaining based at Imperial due to Astbury's inadequate biochemical facilities at Leeds. It also marked the beginning of a lasting friendship, the strength of which, according to Bailey, was rooted in the two men having shared a very similar background. Like Astbury, Bailey had also grown up near Stoke, won his place at university as a result of having earned scholarships, and developed a love of music, in particular Beethoven, one of Astbury's favourites. That it was Bailey who gave the address at Astbury's funeral in 1961 is testament to the depth of friendship that had eventually developed between them.

Two years after Astbury's death, Bailey committed suicide, having suffered recurring attacks of severe depression. Chibnall recalled that, shortly before his tragic death, Bailey had looked tired and haunted but that mention of Astbury and the days at Imperial College brought back 'that old boyish expression of enthusiasm and of eyes that could laugh'.[36] And well they might. For from the work done by Astbury and Bailey on the seed protein edestin, there emerged a powerful idea that was to have far-reaching consequences.

It would be something of an exaggeration to place Astbury's ideas on a par with those of Sir Isaac Newton. But what both men shared in common is that (albeit apocryphally in Newton's case) each found inspiration for their scientific work in the commonplace and edible. For Newton, it was the legendary falling apple; for Astbury, it was the poaching of an egg. By using X-ray diffraction to observe the structural changes that occurred when egg albumin protein was heated, Astbury concluded that, on heating, the protein changed from a soluble, globular form to an insoluble, fibrous form.[37] In its globular form, the protein chain was folded up into an intricate configuration, which was disrupted by heat, causing the chain to unravel and assume the linear, fibrous form. It was this underlying structural change that accounted for the visible changes that occurred when boiling or frying an egg—the change in the albumin from a clear, viscous liquid to a thick, rubbery white solid.

This change from a globular to a fibrous form also explained why enzymes such as pepsin or lysozyme lost their ability to catalyse chemical reactions when they were heated. Above a certain temperature, the thermal energy causes the delicate globular conformation into which the long chain of amino acids has been folded to unravel, leaving the chain in an extended form, rather as if an elaborate piece of origami had been pulled violently apart and the intricate folds smoothed out, leaving just a flat piece of paper. In this extended conformation, the protein chain of the enzyme no longer has the precise three-dimensional shape that allows it to recognise and bind to its target substrate. This explained why organisms such as mammals need to keep their body temperatures regulated at about 37 degrees Celsius—this being the optimal temperature to preserve the fragile folding patterns of mammalian enzymes. Once again, it was the precise shape and structure of the macromolecules in living systems that was shown to be the all-important factor.

Astbury had hoped initially that Bailey's edestin crystals might give him a foothold in the field of analysing globular protein structure using X-rays. But when Bernal had taken his beautiful diffraction patterns of crystalline globular pepsin, Astbury was forced to admit that, once again, he would have to concede this ground to Bernal. But edestin presented another possibility that intrigued Astbury: could the protein chain of globular, crystalline edestin be deliberately unfolded by chemical treatment and then refolded into an insoluble, fibrous form, as Bernal had observed to happen for pepsin?

Ultracentrifugation studies showed that the protein chain of edestin folded up roughly into a sphere, giving it a globular, soluble form. On treatment with the chemical agent urea, however, Bailey found that the globular edestin could be transformed into an insoluble form that gave X-ray diffraction patterns almost identical to those which Astbury had obtained for beta-keratin in wool.[38] Addition of the urea had caused exactly the same transformation as when Astbury had heated egg albumin protein: the intricate folding of the protein chain in the globular form had been disentangled and caused to refold into a stretched, fibrous form.

What Bailey had done with edestin was in principle the very same process by which the soluble globular silk proteins produced in the spinnerets of spiders and certain caterpillars can be secreted as insoluble fibres of incredibly high strength for building webs. And if several million years of evolution in spiders and caterpillars could put denaturation of a globular protein to such effective use, then why might not humans do so as well—and in far less time? In anticipation of the possibility that their work might have some practical use, Astbury, Chibnall, and Bailey filed a patent on the process of unravelling globular seed proteins by chemical treatment and refolding them into a fibrous form.[39]

In the meantime, this work on the denaturation of seed proteins marked the limit of Astbury's work on globular proteins. When Bernal announced that he had obtained X-ray diffraction patterns from crystalline, as opposed to fibrous, pepsin, Astbury had to acknowledge that research into protein structure was now clearly divided into two camps—his at Leeds on insoluble, fibrous proteins, and Bernal's at Cambridge on soluble, globular ones. It was an arrangement to which he seemed able to accommodate himself, saying to Bernal:

I feel that if you and I do not make the most of biological crystallography, we should have our respective bottoms kicked—and if we do make the most of it, then I think things are bound to work out well in the end, in spite of the machinations of nuclear physicists.[40]

By the mid-1930s, Astbury's work on keratin had caught the attention not only of fellow scientists, but also the wider news media. This was in no small way due to the ease with which he could explain his work and the scientific ideas that lay behind it. In particular, Astbury had a talent for being able to take a scientific concept that was very remote from most people's lives, such as the change in conformation of a molecule, and relate it to a familiar, tangible effect. A good example of this skill was in his work on the structural changes of keratin in hair.

His X-ray studies on keratin had shown that their chains could form links, or cross-bridges, between each other and that, when these cross-bridges were broken by means such as heating, exposure to radiation, or treatment with certain chemicals, the keratin chain could be compressed into a conformation that was even more tightly compacted than the alpha form. Astbury called this the 'supercontracted' form and, as an example of its significance, pointed out that it explained why hair could be styled into a 'permanent wave', or 'perm'.[41] One newspaper reported that in a talk given by Astbury at the Royal Institution, '400 wives and sweethearts of Britain's most famous scientists learned why a permanent wave does not come out on wet days', for which he was hailed as 'the lion of the ladies'.[42] Addressing the International Congress of Medicine held in London in 1936, Astbury joked that his work on the molecular structure of keratin in hair might have profound implications for the future of hairdressing:

> There is no longer any need to go to the barber when your hair needs cutting. Expose yourself to X-rays for six hours and then for two hours under steam. The hair will contract to two-thirds its original length.[43]

Communicating his work like this to a wider audience than simply fellow scientists was a role that he relished and in which he excelled.[44] As national newspapers began to report on his work, his research became known well beyond academia and he was soon established as the leading figure in protein fibre research.[45–47] But it could not last. Others were becoming interested in protein structure, and they were very keen to make their presence felt.

5

'The X-Ray Vatican'

Astbury may have lost the soluble, globular proteins to Bernal, but his investigations of the transformation of keratin had established his scientific reputation and brought his work to the attention of the nation. In 1938, his former mentor, Sir William Bragg, put Astbury's name forward for membership of the Royal Society, and two years later Astbury was finally elected as a fellow—one of the most prestigious honours in British science.

Nor did his fame remain confined to Britain. His work fired the imagination of the international scientific community, particularly the eminent US chemist Linus Pauling, who would later become Watson and Crick's main rival in the quest to solve the structure of DNA. Based at the California Institute of Technology (Caltech) in Pasadena, Pauling later became famous for being one of the very few people to win two Nobel Prizes—one for chemistry and the other for promoting the cause of peace—as well as being an advocate of taking large doses of vitamin C as a panacea against illness. However, even in the 1930s his renown had already spread beyond academia, giving him a celebrity status in the wider media.

Born in Oregon in 1901, Pauling developed a fascination with chemistry as a boy, and conducted his own experiments using chemicals 'borrowed' from an old laboratory on the site of the former Oregon Iron and Steel company where his grandfather was a night watchman.[1] Having graduated in chemistry from the Oregon Agricultural College, he moved to California in 1922 to undertake graduate work on the chemistry of simple crystals. What interested Pauling in particular was how different atoms could bond with each other to form molecules. By applying the methods of quantum mechanics to interatomic bonding, Pauling created a set of rules that described how atoms can come together to form molecules. This research was published in a series of papers starting in 1931 and was quickly recognized by the scientific establishment as a major piece of work. In recognition of Pauling's achievements, he was given a full professorship aged only 30 and was awarded the prestigious Langmuir Prize from the American Chemical Society for 'the most noteworthy work in pure science done by a man under 30 years of age'. Albert Einstein showed great interest in discussing Pauling's ideas about chemical bonding with him and the *New York Times* already hailed him as a possible Nobel Prize winner.[2]

The culmination of Pauling's work on bonding was the 1939 publication of his book *The Nature of the Chemical Bond and the Structure of Molecules and Crystals*, which has since become one of the main texts of modern chemistry. But, by the mid-1930s, he had already become intellectually restless and wanted a new challenge. Like Astbury, Pauling was drawn to biology—particularly to the structure of proteins. But where Astbury had used X-ray diffraction to probe the overall structure of protein chains, revealing them to be long chains of amino acids, Pauling adopted a 'bottom-up' approach by exploring how chemical bonding explained the way in which each individual amino acid joined together with its neighbours to form this chain. Like Astbury, Pauling won funding from the Rockefeller Foundation and began a programme of research around trying to understand protein chains by starting with their individual component amino acids and the chemical bonds between them.

Astbury's measurements of the dimensions of the protein chain in keratin were invaluable to Pauling in starting his research, and, while he agreed with Astbury that the keratin chain could adopt two configurations—a compacted alpha form and an extended beta form—he felt that Astbury had missed something important.[3] In Astbury's model of alpha-keratin, the protein chain was compressed in two dimensions like a folded ribbon, but Pauling's knowledge of the angles formed between adjacent amino acids that were chemically bonded together to form a polypeptide chain told him that this was a physical impossibility.

When Astbury was invited to give a lecture tour of the USA in 1937, he visited Pauling to discuss with him the structure of the keratin chain.[4] Pauling was fascinated by Astbury's data and began to apply his knowledge of quantum mechanics and the physical constraints that it placed on the bonds between individual amino acids to Astbury's model of how the keratin chain folded. This discussion of keratin was the beginning of a lifelong correspondence between Astbury and Pauling and also the start of a problem that came to preoccupy Pauling over the years that followed. Convinced that Astbury's model was incomplete, Pauling finally found a solution in 1950, for which, four years later, he was awarded the Nobel Prize.

Being invited to give a lecture tour of the USA was a clear sign of the credibility of Astbury's work. Another was that new sources of financial support now became forthcoming. Having been initially funded by money from the Worshipful Company of Clothworkers, in 1934 Astbury won a £600 grant from the Rockefeller Foundation,[5] established in 1913 when the oil tycoon John D. Rockefeller had faced a major problem. As a result of containing high levels of sulphurous compounds, his crude oil stank of rotten eggs, which had earned it the unenviable name of 'skunk oil',[6] which, unsurprisingly, no customers wanted. Yet, working in their own time, some of Rockefeller's chemical engineers developed a new refining process that not only solved the problem of the smell, but also gave a significantly improved yield of gasoline. Rockefeller was so impressed at the power of fundamental chemistry to solve practical problems that he established the foundation with the aim of using science to further the public good. The foundation survived the economic crash of 1929 better than most other charitable bodies and it is estimated that between the years 1932 and 1959 it donated as much as $90 million in grants to basic scientific research.[7] Much of this research was oriented along a particular theme—the application of physics and chemistry to problems in medicine and biology. This research programme was largely the work of Warren Weaver, the foundation's Director of Life Sciences, and when Weaver and his colleague W. E. Tisdale visited the textile physics laboratory at Leeds in 1935, they were very impressed with Astbury's work and felt that it captured the aims of the foundation perfectly. This marked the beginning of a long-standing and fruitful relationship with the Rockefeller Foundation.

With this generous level of support, Astbury was able to extend the range of his studies to include other keratin-based fibres such as reptile scales, horn, and seagull feathers collected while on holiday at Filey on the East Yorkshire coast. At first sight, this might seem like a random selection of specimens with little to unite them in common, but in a letter written to Bernal in 1930, Astbury hinted that this apparent diversity might only be superficial:

> . . . let me tell you at once of the really exciting discovery I have made. Last Saturday I took a photo of a common fishing float, and it is identical with that of wool and hair, however fine!!! The fishing float is, of course, the quill of a porcupine, but isn't it staggering that such a large epidermal structure—anything up to a foot long—should give an indistinguishable X-ray photo? The implications seem to me to be very great [emphasis in original].[8]

The implications were indeed very great and, in 1938, thanks to a further grant of £10,000 from the Rockefeller Foundation, he was able to explore just what they might be. Contrary to its detractors who damned it as dull and lifeless, Astbury praised the humble wool fibre, saying that with it he had been granted him 'a glimpse of the loom on which the web of life is woven'.[9] With this new level of financial support, he was now able to expand his X-ray studies beyond keratin to encompass other types of protein fibre that might also be part of this web.[10]

One of these was myosin, one of the major protein components of muscle tissue, and Astbury wondered whether his work on the keratin proteins in wool might also be used to explain the fundamental property of muscles—their ability to contract. Could the chains of myosin proteins undergo a similar change in folding that transformed them from a compacted form to an extended configuration, as happened for keratin proteins in wool and hair fibres? At the time, it was impossible to study live muscle tissue using X-rays, but with his research assistant, Sylvia Dickinson, Astbury isolated fibres of myosin from muscles and subjected them to analysis by X-ray diffraction. As predicted, the resulting patterns showed that the protein chains of myosin fibres underwent a transformation in folding from a compacted alpha to an elongated beta form as occurred with keratin in wool.[11–14] But what intrigued Astbury was that, in its compacted form, myosin resembled keratin fibres in which the cross-links between adjacent chains had been broken. This allowed the chains to assume what he called a 'supercontracted' form, which he proposed might be crucial to the mechanism of muscle contraction.[15]

Other subjects included the epidermal proteins such as horns, scales, tortoise shells, and fibrinogen, the main protein involved in the clotting of blood.[16] Yet, despite their apparent diversity, what all these materials shared in common was that, like keratin in wool, their polypeptide chains could change in shape from an alpha to a beta form.

On the basis of this transformation, Astbury termed these proteins the keratin–myosin–epidermis–fibrinogen (k-m-e-f) group, but the real significance of this work is that from it emerged the ideas that would dominate his thinking: that the fibrous chain molecule was a unifying structural motif throughout nature and that living systems could best be understood in terms of these chain molecules. More importantly, from the work on myosin chains in muscle came the idea that it was the transformations in the shape and folding of these giant protein chains that were at the heart of biology: if a chain molecule was important, then it must exhibit some kind of dynamic property and show *movement*.

Astbury's insights into how the analysis of molecular structure could reveal evolutionary relationships were pivotal to the field of molecular biology. With his background in physics, Astbury craved to find a deeper unity and order lying beneath the superficial messiness of biology. So too, did fellow physicist-turned-biologist Francis Crick who, by 1957, believed that he had found just that. In a lecture given as part of a symposium held by the Society for Experimental Biology at University College, London he proposed that it might be possible to trace evolutionary relationships between organisms at the molecular level by comparing similarities in the amino acid sequences of their proteins:

> Biologists should realize that before long we shall have a subject which might be called 'protein taxonomy'—the study of the amino acid sequences of proteins of an organism and the comparison of them between species. It can be argued that these sequences are the most delicate expression possible of the phenotype of an organism and that vast amounts of evolutionary information may be hidden away within them.[17]

Today, this approach has confirmed the idea first proposed by the Victorian scientist Thomas Huxley of an evolutionary link between birds and reptiles. But Crick was not actually the first to suggest that evolutionary ancestry might be evident at the molecular level. Nearly three decades before Crick gave his lecture proposing the idea of 'protein taxonomy', Astbury had already made a very similar suggestion—albeit one based on a different principle to that of Crick. Unlike Crick, Astbury had no means of comparing amino acid sequences in different proteins but, having extended his X-ray studies from keratin fibres in wool to those found in the scales of reptiles and feathers of birds, Astbury had found striking similarities in the three-dimensional structure of these molecules, which he said was likely to be of 'of peculiar significance for palaeontology and evolutionary theory in that it is a direct indication by the methods of molecular physics of an affinity between reptiles and birds which differentiates them from the mammals'.[18, 19]

No wonder then that Astbury later praised wool as being 'the most exciting protein in the world', for he believed that from his studies on wool he had glimpsed a principle that promised to unite the apparent diversity of the living world.[20] He now wanted to show that this principle was not just confined to proteins, but that it also extended to other kinds of biological chain molecules, for example, polysaccharides.

Polysaccharides, such as cellulose, are long chains formed by the chemical bonding of smaller sugar molecules. Indeed, it was cellulose, found in the cell wall of plants, that provided the inspiration for Astbury's model of the keratin protein fibre in wool, and it also provided him with the opportunity to expand his X-ray work beyond proteins to include polysaccharides.

Thanks to the young scientist Reginald D. Preston, Astbury got his chance to expand his work beyond proteins. The son of a local self-employed builder who also happened to be an avid reader of the journal *Nature*, Preston had inherited his father's love of learning and could read Milton's 'Paradise Lost' by the time he was four years old.[21] As a result, his reading was superior to that of most of the teachers in his school in Wortley and, thanks to this academic ability, Preston won a series of scholarships throughout his school career, where he ended up at the University of Leeds Physics Department. Towards the end of his studies, Preston got the chance to do some work in the Department of Botany. Astbury's passion for bringing the methods of physics to bear on biological problems was gaining favour, and the botanists at Leeds were keen to recruit a physicist to help them study fibrous materials in the cell walls of plants and algae. This proved to be a field in which Preston excelled himself and one in which, thanks in part to Astbury's influence, he became an eminent expert on biological fibres—such that, according to one obituary, his skills were called upon to analyse fibres in the Turin Shroud.[22]

It was during a chance encounter in the staff cloakroom that Preston first met Astbury, and, in the conversation that followed, Astbury

became excited to learn that Preston had obtained crystals of cellulose that would be ideal material for examination by X-ray analysis—
although their structure had been studied by X-ray analysis, no-one
had yet used this method to observe these fibres in their native cellular environment. The source of the cellulose fibres was the organism
Valonia ventricosa, or 'sailor's eyeball', an alga whose single cells were giant spheres that could be up to 2 cm in diameter and covered with an
intriguing striated pattern visible when viewed down a microscope.

Astbury granted Preston free use of his X-ray facilities, but when their
first paper appeared, Preston was surprised to find his name had been
omitted. In response, Astbury offered the rather dubious justification
that four authors were enough, and that Preston was currently too junior to merit inclusion. Despite the exclusion, they went on to enjoy
a fruitful scientific collaboration that soon evolved into a friendship,
which, as Preston recalled, was helped largely through a shared passion
for classical music:

> as with science, Astbury was happiest sharing his music and nothing
> would do but that we should play piano duets together. So I bought
> the appropriate scores of Beethoven and Mozart and Haydn and Bach—
> Astbury had them already—and I practised and practised the secondo
> parts both solo and alone and in duet with Astbury. We must have played
> the Beethoven symphonies dozens and dozens of times. Astbury was a
> high-class performer and the way he played and the way he made me
> practise at one time led me almost to the illusion that I wasn't doing
> too badly either; at least, if I listened to him and ignored my playing it
> sounded alright. These duets of ours became the feature of many house
> parties Astbury gave during these years, replacing the earlier playing of
> the real thing on his beloved gramophone records. We usually ended up
> with Astbury stripped to his shirt, banging the piano and saying some-
> thing like 'Preston, it's no good. You're playing too fast and I can't keep
> up with you', whereas in truth I had been skipping bars for the last ten
> minutes and at the moment was playing at least five bars behind.[23]

What Astbury and Preston lacked for harmony when playing a duet on
the piano they more than made up for in the laboratory. Their X-ray
analysis of *Valonia* revealed that the enigmatic striated patterns visible
under the microscope arose from cellulose fibres being aligned in two
different orientations. While one set of cellulose fibres were wrapped
around the circumference of the cell in a series of meridians, the other

was wound around the surface of the cell in a twisting left-handed spiral. The same pattern was found in another alga, *Cladophora*, leading Astbury and Preston to speculate that spiral growth of cellulose fibres might be a feature not only of algae, but also of cell walls throughout the higher plants.[24] Moreover, they proposed that within the impressive mathematical regularity of these spirals was a clue to a deeper underlying order in biology that dictated the growth of cells and whole organisms. Once again, studying the shapes of large chain molecules offered a glimpse of common underlying patterns that spoke of a deeper unity in the living world.

And Astbury's love of classical music offered him a powerful metaphor for his other great passion of molecular biology:[25]

> I should like to say again, that in the symphony of creation the greatest instrument of all, the chosen instrument on which nature has played so many incomparable themes, and countless variations and harmonies, is the chain-molecule ... Or may I put it even more profoundly and say simply that we are all mostly fibres and water ... with sometimes a little alcohol.

To some of his critics, the breadth of Astbury's research programme was a fundamental flaw and weakness in his work. In their view, it lacked focus and was too eclectic and diverse to deliver results of any real substance. As colleague Ian MacArthur later remarked:

> He brought his findings to market in the green ear, but would not clear the weeds nor suffer the system and technique necessary for the harvest.[26]

In the face of such criticisms, Astbury preferred to justify his approach in the words of his favourite fictional detective, Lemmy Caution:

> You gotta realise that in a job like this you gotta start something. It is no good just hangin' around an' looking for clues. The great thing to do is to throw a spanner into the works an' get everybody sorta annoyed with each other.[27]

Acting on Caution's advice, Astbury threw spanners in all directions, and it proved to be a successful strategy. By examining a broad range of biological materials with his X-ray methods, he showed repeatedly that common structural motifs are used in Nature at the molecular level to generate a stunning repertoire of forms and structures. Nature has, he said:

built up a textile business that far transcends any man-made effort, and really the only thing new that has come out of modern textile science, itself one of the triumphs of the century, is the discovery of how very old it all is. Fibres, visible and invisible, are the principal structural components of all biological tissues, and the fibre concept is as old as life.[28]

Astbury's initial fears that a move to Leeds would mean heading into a scientific wilderness were now well and truly vanquished. In recognition of his reputation as the world leader in X-ray diffraction studies of biological fibres, the Austrian scientist Max Perutz, who would later win a Nobel Prize for elucidating the structure of haemoglobin, once hailed Astbury's laboratory at Leeds as 'The X-ray Vatican'.[29] News of his work had spread beyond academia into industry and the national media. And at the heart of it all lay keratin, the fibrous protein of wool and hair that had once been described as 'thoroughly dead, unbelievably dull and unprofitable scientifically, and altogether the kind of protein . . . that no respectable, aspiring biochemist would touch with a barge pole'.[30]

From his X-ray observations of keratin in wool was born an idea that came to define Astbury's entire research career: that biological function is dictated by shape. It was this idea on which all his success as a scientist had been built. But he also felt that his work on keratin had given him a glimpse into another, perhaps more fundamental, question in biology, which is how biological systems replicate to make new copies of themselves.

Astbury speculated that a clue to this question might lie in the very fact that proteins were long chains of amino acids. It was possible, he suggested, that one protein chain might act as a template for the assembly of a second chain, with the ordered arrangement of amino acids in the parent chain directing assembly of the daughter chain.[31] He proposed several models for how two proteins chains might interact to make this possible, in one of which the parent and daughter chain ran in opposite directions. This gave rise to a pattern of symmetry that in the jargon of the X-ray crystallographer was known as 'C2-type' symmetry. Had Rosalind Franklin been familiar with Astbury's model, then she might well have spotted from her X-ray work that 'A'-form DNA also gave C2 symmetry, suggesting that it consisted of two chains running in opposite directions. This, along with knowing how the bases on these two chains pair up, would have constituted the final two steps that would have enabled her to solve the structure of DNA ahead of Watson and Crick.[32]

What Astbury did not realise at the time that he was proposing these models was that the secret of replication in living systems lay not with proteins and polysaccharides, but in a third type of biological fibre. It was a white stringy material that was precipitated from the nuclei of cells when they were ruptured with detergent and mixed with ice-cold alcohol. Isolated from the thymus gland of calves, it was then known as 'thymonucleic acid'. Today we know it by a far more familiar name: deoxyribonucleic acid, or DNA.

Astbury's ignorance of DNA did not last long. From his work on protein fibres, he had conceived the powerful idea that biological function could be explained in terms of molecular shape. Thanks to his work with Preston, he had expanded this idea to include a second class of biological fibres: the polysaccharides. Could the idea now be extended even further to also explain the function of this third class of biological fibres, the nucleic acids? Astbury was convinced not only that it could, but also that doing so would reveal a single principle that unified the whole of biology.

6

'A Pile of Pennies'

Watson and Crick may have uncovered its molecular structure, but DNA itself was discovered nearly 100 years earlier when a young Swiss physiologist by the name of Friedrich Miescher isolated some stringy white fibres from an extract of cells. The essential process that Miescher used to extract DNA can easily be repeated in the kitchen with reagents no more sophisticated than strawberries, washing-up liquid, table salt, and ice-cold rum. Miescher's own source of DNA, however, was far more unsavoury than soft fruit.

After graduating from medical school in 1869, Miescher was trying to find a direction for his future career. On account of suffering from a severe hearing impairment, however, he felt that he was not best suited for interacting with patients. As a consequence, most areas of medical practice were not an option for him, and he chose instead to pursue medical research. Later that year, he went to Tübingen, Germany, where he hoped to pursue his interest in studying the contents of tissues and cells. Here, his new supervisor, Professor Felix Hoppe-Seyler, gave him the task of characterising white blood cells found in the lymph glands. But Miescher soon faced a major problem. It seemed to be impossible to obtain enough white blood cells from the lymph glands in quantities sufficient for his experiments. Without a ready supply of white blood cells in large quantities, his work was at a dead end, and so he was forced to look around for an alternative source of these cells.

To his relief, he found one soon enough, albeit from a rather unlikely supplier. The bandages discarded from surgical operations on patients at a nearby military clinic were full of pus that was rich in white blood cells. Collecting these bandages provided Miescher with a steady and abundant supply of white blood cells, and he set to work trying to analyse their chemical composition. First, he washed the pus from the bandages with a mild solution of sodium sulphate before treating them with warm alcohol to remove fats. Next, he separated the cell nucleus from the rest of the cell by carefully adding dilute hydrochloric acid to the cell extract. The result was a grey, solid mass that contained mostly cell nuclei. Miescher treated this precipitate with the digestive enzyme pepsin to remove any lingering traces of cytoplasmic protein. Finally, on addition of mild alkali, a white fluffy substance was seen to precipitate from the nuclei.

It was this material in particular that intrigued Miescher. Chemical analyses showed that it was mildly acidic and contained carbon, nitrogen, hydrogen, and oxygen just like a protein. But, unlike a protein, it was also rich in the element phosphorus and was found only in the nuclei of the white blood cells, which led Miescher to name it 'nuclein'.

Miescher's supervisor was cautious about the interpretation of these results and insisted that the chemical analyses be repeated, which delayed publication of this work until 1871. By this time, Miescher had returned to his home city of Basel, where he had found a second abundant source of nuclein. The particular section of the River Rhine that flowed past Basel was a spawning ground for millions of salmon, whose sperm cells bulged with giant nuclei that occupied 90% of their volume. Using nuclein extracted from these salmon sperm cell nuclei, Miescher continued his research under working conditions that were far from ideal. Rather than being given his own laboratory, he was restricted to performing his extractions of nuclein in one small corner of a larger chemistry laboratory, while the chemical analysis of the substance had to be performed in a corridor in a separate building. The entire process was laborious and time consuming, as Miescher himself described:

> When nucleic acid is to be prepared, I go at five o'clock in the morning to the laboratory and work in an unheated room. No solution can stand for more than five minutes, no precipitate more than one hour before being placed under absolute alcohol. Often it goes until late in the night. Only in this way do I finally get products of constant phosphorus composition.[1]

The harshness of this regime took its toll. Having suffered a steady breakdown in his health over the years, Miescher succumbed in 1895 to tuberculosis and died. But his research student Robert Altmann continued his work and made a more detailed chemical analysis of nuclein, showing that it actually had two very distinct chemical components. One of these was rich in protein, while the other was a mild acid, which Altmann called 'nucleic acid'.

For about the next 29 years until 1900, Altmann and his colleague Albrecht Kossel made exhaustive analyses of the nucleic acid component of nuclein and showed that it was composed of four smaller chemicals called bases—adenine, cytosine, guanine, and thymine. But while Altmann and Kossel's analysis had determined the chemical composition of nucleic acid, it gave no insight as to how these four chemicals might be joined together to form a single molecule of nucleic acid. It was rather like showing that a building is made of four different types of brick but with no idea of the shape in which the bricks are arranged to actually form the building. The question has certain parallels with the debate in Chapter 5 about how individual amino acids came together to form proteins. Did the four bases chemically bond to each other to form small units, or were they linked together to form giant chains?

A young Jewish Russian chemist, Phoebus Aaron Levene, believed that he had an answer. Levene had graduated from the Imperial Medical Academy in St Petersburg, where he had been taught by the physiologist Ivan Pavlov (later to become famous for his eponymous dogs that salivated in response to hearing the ringing of a bell) and the composer and chemist Alexander Borodin. In 1891, Levene moved with his family to the USA to escape the persecution of Jews in Russia. He began working on nucleic acids in 1896 at the Pathological Institute of the New York State Hospitals, before eventually becoming director of the division of chemistry at the Rockefeller Institute for Medical Research. Here he ran a massive research team where the duties of his staff, in addition to analysing the composition of nucleic acids, included chauffeuring him in his own car to the opera. One of his 'chauffeurs', Stuart Tipson, recalled Levene's approach to work thus:

> He would get up at 6 o'clock in the morning, get washed and shaved. Then he would read a French novel, and go out for a walk, and then come in for breakfast. Then he would take a fifteen-minute nap and he would go to work at the Institute.[2]

The nucleic acids that Levene used for his work came from two different sources. The first of these was isolated from yeast and is what we now call ribonucleic acid (RNA). In 1909, Levene proposed that this 'yeast nucleic acid' was built up of smaller chemical units called nucleotides. Each nucleotide contained one of four possible bases—adenine, guanine, cytosine, or uracil, which were chemically bonded to a phosphate group and the sugar ribose. The sugar and phosphate groups of each nucleotide acted as chemical bridges, forming bonds from one nucleotide to the next and so in turn linking up individual nucleotides to form a giant chain.

The other main nucleic acid that Levene used in his work was known as 'thymonucleic acid' on account of being extracted from the thymus gland of calves. But chemical analysis of thymonucleic acid proved to be much more challenging and difficult than had been the case for 'yeast nucleic acid'. The chemical conditions used to break the bonds between individual nucleotides had to be very carefully controlled to avoid reactions that altered the sugar and bases, giving rise to erroneous results. Each of the four different bases required a different set of conditions to be crystallised so that their total mass could be measured, and these conditions had to be painstakingly determined by trial and error. As a result, it took Levene nearly another 20 years before he finally managed to successfully identify the components in the nucleotides of thymonucleic acid as being the four bases adenine, guanine, cytosine, and thymine, and the sugar deoxyribose, from which we derive our modern name for 'thymonucleic acid'—deoxyribonucleic acid, or DNA. By this time, Levene had also proposed the first theoretical model for how these individual components might link together to form a bigger structure. From largely theoretical calculations, Levene had concluded that the proportions of adenine, guanine, cytosine, and thymine in a molecule of thymonucleic acid were roughly equal and that this gave him an important clue to the structure of a nucleic acid chain. In what came to be known as the 'tetranucleotide hypothesis', Levene proposed that thymonucleic acid was a long chain formed by smaller regular repeating units composed of all four nucleotides, chemically bonded together.[3] Despite being based largely on theory, the 'tetranucleotide hypothesis' was widely accepted as a plausible structural model for what a nucleic acid chain might look like.

While Levene had made great strides in determining the chemical composition of nucleic acids, his tetranucleotide hypothesis gave no clue as to what they did, and their function in the cell remained a mystery. When he had first isolated 'nuclein' back in 1869, Miescher had speculated that it might be important in cellular metabolism as a store for phosphorus, but by the early twentieth century biologists were beginning to suspect that they might have a far more interesting role. This was prompted by the observation that nucleic acids were one of the main components of the two most basic structures in biology capable of replication: chromosomes and viruses.

Chromosomes can be observed as threadlike structures that become visible during cell division when viewed through a microscope. Work by geneticists in the first few decades of the twentieth century on mutations in the fruit fly *Drosophila melanogaster* had shown that inheritable traits such as red and white eyes could be mapped to specific physical regions on a particular chromosome, suggesting that this region of the chromosome somehow controlled inheritance of eye colour. Moreover, by observing the particular patterns and ratios in which physical traits were inherited, the geneticists could deduce which traits mapped to the same chromosomes and the relative distance between regions of the chromosome controlling the inheritance of these traits.

While geneticists studied the patterns in which particular traits were inherited and how they mapped to chromosomes, cell biologists studied the physical properties of the chromosomes. Chemical analysis showed that the chromosomes were composed of both proteins and nucleic acids bound together in a complex. When stained with dyes and viewed under the microscope, the chromosomes appeared to consist of a number of striped bands, some of which correlated with regions responsible for particular inherited traits studied by the geneticists. This raised the tantalising possibility that there was some direct relationship between an inherited physical trait and a chemically distinct region on a chromosome.

At the heart of such a relationship was the idea of the gene. Coined in 1909 by the Danish biologist Wilhelm Johannsen, the term 'gene' was used by geneticists to refer to the specific hereditary traits, for example, eye colour in *Drosophila*. However, at the time it was unclear whether or not the gene was an actual physical entity. In rather the same way that an interest rate has no physical existence, and is an abstract (but nonetheless essential) tool to an economist, the gene was an abstract concept used by geneticists to obtain meaningful patterns from the numerical data they obtained when analysing the progeny from crosses between fruit flies. Now, however, the work of the cell biologists began to suggest that the gene might well be more than an abstraction—that there might be a physical basis to the mechanism of heredity and that this was rooted in the material composition of chromosomes.

It was an idea that gained support from yet another field of biology— the study of viruses—where research had focused on one virus in particular, the tobacco mosaic virus (TMV). Tobacco plants infected with TMV develop variegated leaves and stunted growth, which reduces their yield considerably. As a result, research into TMV was considered to be of great value to the US economy. In 1935, the American virologist Wendell Meredith Stanley, working at the Rockefeller Institute, isolated a crystalline protein from the juice of leaves on Turkish Tobacco plants infected with TMV. Stanley showed that the protein crystals had exactly the same infective properties as TMV and so concluded that they were the active agent of TMV that allowed it to infect plants and replicate itself within their cells.[4] The idea that the entire pathology and behaviour of a virus might all reside in a single protein was a momentous discovery and one that was a triumphant vindication of the Rockefeller Institute's grand research programme of reframing biological problems in terms of physics and chemistry.

Astbury hailed Stanley's discovery of the crystalline TMV protein as 'undoubtedly the event of the century'.[5] Shortly afterwards, more detailed chemical analysis by the British scientists Bill Bawden and Fred Pirie at the Rothamsted Agricultural Research Centre showed that the material isolated by Stanley was made not only of protein, but also contained nucleic acid.[6] Like the chromosomes, the infectious virus particle was therefore considered to be what became known as a 'nucleoprotein'.

But the question still remained—what exactly was the role of the nucleic acids in these nucleoproteins found in viruses and chromosomes? Thanks to a chemical staining technique called the Feulgen method, the amount of nucleic acid in a cell could now be quantified and was shown to increase significantly during cell division. Although this was strong evidence that nucleic acids had some role in biological replication, it was still unclear exactly what their function was in this process. The observation that the amount of nucleic acid increased during cell division was open to interpretation: it might well mean that nucleic acids were the central molecular agent in enabling chromosomes and viruses to copy themselves, or that replication might equally be controlled by some other agent in the course of which amounts of nucleic acid increased as a secondary effect.

While the chemical nature of the genetic molecule remained unclear, what was certain was that it must possess one key property—a high degree of variation in its structure. This was a point made clear in a series of lectures delivered in 1943 by the physicist Erwin Schrödinger.

Better known for proposing a famous thought experiment involving an unfortunate cat to illustrate some philosophical problems raised by quantum mechanics, Schrödinger was awarded the 1933 Nobel Prize in Physics for his contribution to this field. But this honour counted for little when the Nazis took control of his native Austria in 1938, and so Schrödinger fled to Dublin.

It was while living in exile in Dublin that Schrödinger delivered a short course of public lectures under the title 'What is Life?' in which he attempted to understand living systems using the methods of a physicist. The following year, these lectures were published as a short book that has since been hailed as a kind of 'Uncle Tom's Cabin'[7] of molecular biology—a reference to the 1852 novel by Harriet Beecher Stowe that many writers have argued awoke the American conscience to the immorality of the slave trade and helped the abolitionist cause in the build-up to the Civil War of 1861–1865. For many scientists, Schrödinger's book has come to be seen as a call-to-arms to tackle biological questions using the methods of physics and chemistry, and both James Watson and Francis Crick acknowledged that Schrödinger's book was a key influence on their thinking.[8] One of the main questions Schrödinger asked was how organisms were able to pass on physical traits from one generation to the next.[9] Schrödinger wondered how it could be that successive members of the Habsburg Royal Family all shared such a distinctively shaped lip that seemed to pass unchanged down the generations? By way of an answer, he speculated that the transmission of physical traits from one generation to the next might occur at the molecular level. It might be possible, he argued, that inheritable biological traits could somehow be represented by variation in the structure of a sufficiently large molecule. Terming such a molecule a 'code-script', he then went on to identify the two key properties that a suitable molecule must have to fulfil this function. Firstly, it should be thermodynamically stable and not degrade easily. This could be achieved if the molecule were a long chain of smaller molecules chemically bonded together. Secondly, in order to encode all the information needed to make something as complex as an oak tree, an elephant, or a Habsburg lip, this long chain must have the capacity for immense structural variation. To understand why this should be the case, it may be helpful to consider how an alphabet works as an information-carrying system, or 'code-script'. Using the Western alphabet of 26 letters, we are able to generate a vast amount of words, which in turn can be combined into sentences to transmit an even greater amount of information. Yet imagine by contrast how limited our scope for forming words and transmitting information through sentences would be if this alphabet were restricted to only four letters.

Of all the different kinds of macromolecule inside a living cell, it was proteins that seemed the best candidates to satisfy both of Schrödinger's requirements for a genetic molecule. As well as being chemically stable, their chains were made of up to 20 different amino acids and so had the potential to generate the immense amount of structural variation that Schrödinger argued was necessary for carrying biological traits. The nucleic acids, by comparison, seemed to be disappointingly dull and inert. According to Levene's tetranucleotide hypothesis, they were just long chains formed by repetition over and over again of units containing the same four bases. It was difficult to see how such a monotonous molecule could generate anything like the kind of variation seen in proteins.

Yet if proteins were the most likely candidate for being molecular carriers of heredity, why were nucleic acids always so closely associated with proteins in replicating units such as viruses and chromosomes? One answer was that perhaps the nucleic acids served merely as a kind of molecular scaffold acting as physical support—a kind of molecular girder—while the proteins did the important work in heredity. This might explain why they were such a major component of chromosomes and viruses and why their amount increased during cell division. However, chemical analysis alone would be insufficient to answer this question—what was required was a means of probing the physical structure of the nucleic acid molecules.

Inspired by Stanley's isolation of a crystalline virus, at Cambridge Bernal started using X-ray diffraction to study other viruses such as potato X and bacteriophages, which are viruses that specifically infect bacteria. It was to Bernal therefore that Bawden and Pirie turned when they wanted more structural information about the nucleoprotein they had isolated. When Bernal analysed the samples by X-ray diffraction, he was excited to report that the X-ray diffraction pattern hinted at some regular repeating structure not only in the protein component, but also in the nucleic acid. Bernal felt sure that this observation would interest Astbury, but the idea that nucleic acid structure might be probed using X-ray diffraction was not new to Astbury, who wrote in reply:

> I was rather amused at your getting a fibre photograph of nucleic acid.
> I got a similar photograph 3 or 4 years ago from a specimen which I re-
> ceived from Schmidt, and I have been thinking of investigating it in more
> detail now with regard to the chromosome business.[10,11]

Actually, both Astbury and Bernal's attempts to study nucleic acids
using X-rays had already been preceded. In 1920, the scientists R. O.
Herzog and W. Janke working in Germany had already tried to use
X-ray diffraction to study the molecular structure of nucleic acids,
but were unaware that their samples of nucleic acid contained mostly
highly degraded material in which the chemical bonds between the
individual nucleotides in the DNA chains had been broken.[12] Consequently, their DNA samples were mostly just free nucleotides and not
the highly regular ordered chains that give a clear X-ray diffraction pattern. As a result, all they obtained on their photographic plates was an
unintelligible mess.

Amid all the excitement about keratin and protein chain structure, Bernal's news made Astbury wonder whether nucleic acids might
be worth a second look. After all, he had lost the field of globular
protein crystal structure to Bernal and was reluctant to see his old
colleague now claim the nucleic acids solely as his own. A chance for
renewed work on the structure of nucleic acids came thanks to the
work of Swedish scientists Einar Hammarsten and Torbjorn Caspersson. Working at the Karolinska Institute in Stockholm, Hammarsten
and Caspersson had improved the method of extracting and purifying thymonucleic acid so that the final product did not break down.
As a result, they were able to obtain fibres of nucleic acid in which the
giant molecular chains of nucleotides remained undegraded. In 1936,
Caspersson took this high-quality material to Rudolf Signer's laboratory in Bern. Signer was using an optical technique that studied how a
fibre affects the rotation of polarised light. This property, termed 'negative birefringence', was first reported for nucleic acids by Bernal in
his work on TMV, but, with the high-quality samples of thymonucleic acid, Caspersson and Signer were able to further extend this work.
From measurements of the rotation of polarised light, they deduced
that each molecule of thymonucleic acid was a long rod, the length of
which was 300 times its width and which had a molecular weight of between 500,000 and 1000,000. Furthermore, the bases of the nucleotides
must be stacked up on top of each other in a columnar structure like

a fibre.[13] This was the first piece of structural information about DNA, but it was the absolute limit of what could be determined by this optical method—a more detailed study of the nucleic acid fibre required a more powerful technique. X-ray diffraction was the ideal weapon of choice, and to study fibres using this method, only one laboratory in the world at that time had the capability—Astbury's at Leeds.

What Astbury needed in addition to the samples of thymonucleic acid was a skilled X-ray crystallographer to work on them. Thankfully, by the time that the fibres of nucleic acid arrived from Hammarsten, he had already found one. When, in 1937, Lawrence Bragg heard that Astbury was looking for a good X-ray crystallographer, he had written to him saying that he had an 'excellent candidate' for the post working in his own laboratory at the time.[14] Florence Bell (Figure 12a) was a young woman with an impressive range of interests. As well as being able to read French and German, she enjoyed playing the piano—always a sure way to win Astbury's respect.[15] Nor were her talents confined merely to the intellect, for she also enjoyed lacrosse, netball, tennis, and swimming and, as Astbury once remarked, even had an impressive aptitude for cricket.[16] But it was in science that she truly excelled. As Head Girl at Haberdashers' Aske's School in London, Bell had shown great flair for mathematics and science, which had led her to read Natural Sciences at Girton College, Cambridge, where she specialised in chemistry, physics, and mineralogy.[17,18] After completing her degree, she worked briefly as Bernal's research assistant at the Cavendish Laboratory in Cambridge, where she was trained in X-ray crystallography. Afterwards, she left Cambridge to join Lawrence Bragg's group in the Physics Department at the University of Manchester, where, supported by funding from Imperial Chemical Industries, she used X-ray diffraction to study the structure of sterols.[19]

Bell was keen to move to Leeds as she was engaged to a doctor there and the post of research assistant in Astbury's laboratory, funded by the Rockefeller Foundation, was ideal as it was a much longer-term post than her current position with Bragg. In his letter, Bragg pointed out that there was currently a shortage of good crystallographers as industry was 'snapping them up' and said that, while he was sorry to see her go, he could vouch for her 'as somebody who is capable and experienced and really gets on with the job'.[20]

It was a description that Astbury quickly recognised as true and, in acknowledgement of Bell's calibre as a scientist, he chose her to

(a)   (b)

Figure 12 a) Florence Bell (1913–2000) (b) 'Woman Scientist Explains'—
Headline from 'The Yorkshire Evening Press' 23rd March 1939 reporting on
Florence Bell's presentation of research in Astbury's laboratory to delegates at
a conference held by the Institute of Physics in Leeds in 1939.

a) Photograph courtesy of Mr. Chris Sawyer. b) Astbury Papers MS419 Press clippings
book A.1, Special Collections, University of Leeds, Brotherton Library. Reproduced
with the permission of Special Collections, University of Leeds.

give a presentation explaining the work of his laboratory to assembled
delegates at the March 1939 Institute of Physics conference at Leeds.
The press reports covering this meeting are certainly striking for the
insight that they offer into how female scientists were perceived at
the time. In addition to relating the scientific content of her presentation, the newspapers felt that it was necessary to inform their readers
that Bell was 'a slim 25-year-old'. Reports declaring that 'Aged 25 she
will lecture to scientists' seemed to completely overlook the glaring
fact that Bell was indeed herself a scientist![21] Moreover, the implicit
sense of shock, surprise, and confounded expectation in the *Yorkshire
Evening News*' headline 'Woman Scientist Explains' has echoes of Dr
Samuel Johnson's disparaging comments about a dog walking on its
hind legs.[22]

Nor was Astbury himself above holding similar views. When proposing that his former colleague Kathleen Lonsdale, with whom he had

taken his first steps into scientific research at William Bragg's laboratory at the Royal Institution, be put forward for election to the Royal
Society, he wrote to Lawrence Bragg saying:

> I suppose the suggestion was bound to come up sooner or later that
> women should be put up for the Royal Society, and once that is accepted
> I don't think you could find a woman candidate more likely than Mrs.
> Lonsdale to be successful. I should put her at quite the best woman sci-
> entist that I know—but that probably is as far as I am prepared to go,
> because I must confess that I am one of those people that still maintain
> that there is a creative spark in the male that is absent from women,
> even though the latter do so often such marvellously conscientious and
> thorough work after the spark has been struck.[23]

In asserting that women lacked the 'creative spark' required to perform
pioneering science, Astbury was oblivious to his own contradictions.
He respected and valued Bell's intellect and prowess as a scientist enor-
mously, such that he christened her his 'vox diabolica', or devil's advocate,
for her formidable ability to challenge his tendency to let his imagination run ahead of his experiments (Figure 12).[24] There was no greater
testament to this than when he entrusted her with the task of carrying
out X-ray diffraction studies of the DNA samples that he had received
from Hammarsten.

Although Bell's PhD thesis covered a wide range of problems, including determination of the structure of collagens (the other main group
of fibrous proteins) as well as the question of structure in the non-
fibrous (globular) proteins, it was her X-ray studies of nucleic acids that
were to have the biggest impact. These came from a variety of sources,
including yeast, pancreas, and TMV, but it was her data obtained from
the calf thymonucleic acid sent by Hammarsten that yielded the most
information. To the untrained eye, Bell's X-ray photographs might appear to be just a blurred mess of smeared black circles and dots, but they
marked a key moment in the history of science, for they represented the
first successful use of X-ray diffraction to study the structure of DNA—
a journey that would culminate with Rosalind Franklin's 'Photo 51'.
What Astbury and Bell did not realise at the time, however, was that,
depending on its water content, DNA could assume two forms, each
of which gave very different diffraction patterns. These were the two
forms later discovered by Rosalind Franklin, who designated them as

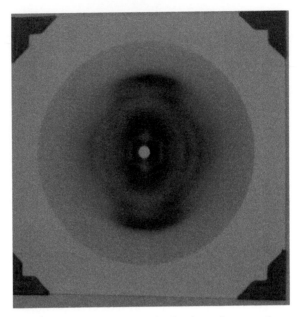

Figure 13 X-ray diffraction photograph of sodium thymonucleate taken by Florence Bell.

From 'X-ray and Related Studies of the Proteins and Nucleic Acids', PhD thesis, University of Leeds, 1939. Reproduced with kind permission of the University of Leeds.

the 'A' and 'B' forms, according to features such as the distance required to make a complete helical turn (the 'pitch'), the alignment of the deoxyribose groups, and the angle of tilt that the base pairs make with the helical axis. Astbury and Bell were unaware that their samples of nucleic acid were a mixture of both the 'A' and 'B' forms and so gave diffraction patterns that were superpositions of the two, yielding only very limited structural information about the molecule (Figure 13).

But while the information from Bell's pictures was limited, this did not mean that they were worthless. On the contrary, by carefully studying the dimensions of the rings and spots in these pictures, Bell and Astbury confirmed that the DNA molecule was composed of a stacked column of nucleotides, which they described as resembling 'a pile of pennies' (Figure 14). This in itself was an important result, but their real achievement was to deduce from the photographs quantitative information about the actual dimensions of the DNA molecule. Knowing that there was a direct mathematical relationship linking the

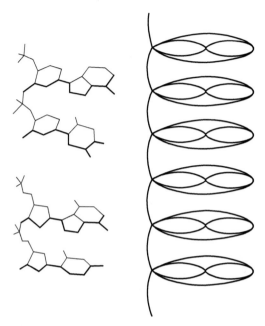

Figure 14 Astbury and Bell's 'pile of pennies' model of the structure of DNA. Published in Astbury, W.T., Bell, F.O., 1938. Some recent developments in the X-ray study of proteins and related structures. *Cold-Spring Harbor Symposia on Quantitative Biology*, 6, pp. 109–118. Reproduced with the kind permission of Cold Spring Harbor Laboratory Press.

spacing of the patterns of rings and spots in Bell's X-ray diffraction pictures and the dimensions of the molecule, they were able to measure the distance between adjacent bases along the length of the DNA chain as 3.4 Angstroms (0.34 nanometres).

Bell later wrote that, with his measurement, she and Astbury were left feeling 'ecstatic'.[25] The reason for their jubilation was that this distance corresponded almost exactly with the spacing between successive amino acids in the chains of keratin proteins that Astbury had studied in wool. Astbury was convinced that this could be no mere coincidence and hinted at a deep and intimate structural relationship between protein and nucleic acid chains.[26] Understanding the details of this relationship promised not only to explain the molecular structure of chromosomes and viruses, but also the one property that made them so special—their ability to replicate. No wonder then that Astbury and

Bell were excited—with one measurement they believed that they had been granted 'a glimpse of one of the most fundamental calibrations in Nature's workshops' and, with it, an answer to one of the most fundamental riddles in biology!'[27]

Astbury's enthusiastic conviction that with this single measurement he had grasped one of Nature's deepest secrets has been memorably described by the historian Robert Olby as having the 'devotion of a youthful mystic'.[28] It was proteins that were, he declared, the 'long scroll upon which the patterns of life were written'.[29] But how were these 'patterns of life' replicated and transmitted? Astbury and Bell believed that they had been given the first hint of an answer, and it lay within the nucleic acids. According to Levene's tetranucleotide model, nucleic acids were nothing more than a dull, inert molecule made up of monotonous repeats. It was difficult to see how such a structure could fulfil a function as complex as replication. It was far more likely that nucleic acids were simply a kind of molecular scaffold offering structural support to proteins or, as it has been more poetically described, 'the wooden stretcher behind the Rembrandt'.[30] But Astbury was more cautious about dismissing the role of nucleic acids so quickly. For him, it was a 'fair conclusion that nucleic acid is essential to the process [of viral replication], and in fact to all processes of biological duplication'.[31–33]

He proposed that replication in biology was dependent on an intimate association between nucleic acid and protein chains, which came about thanks to the identical spacing of the bases and amino acids. During this process, the polypeptide chain of a globular or fibrous protein unravelled and became fully extended to be stretched out along the length of the accompanying nucleic acid chain. The bases protruding from the nucleic acid chain then paired up with the amino acids running along the length of the protein chain rather like the teeth on opposite strands of a zip.[34] The resulting molecular complex would be a hybrid of nucleic acid and protein, or a 'nucleoprotein' in which the nucleic acid might somehow serve to direct and organise replication of the protein chains (Figure 15).[35]

Models, however, are only half the story. What was needed was some more experimental support for this hypothesis and, to obtain this, Bell attempted to show how protein–nucleic acid hybrids could be generated in vitro. Taking crystalline samples of the globular protein clupein, she treated them with hydrochloric acid to denature their polypeptide chains and then mixed them with a viscous solution of sodium

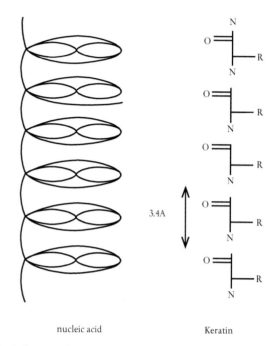

nucleic acid   Keratin

Figure 15 Astbury and Bell were 'ecstatic' that the distance of 3.4 Angstroms between successive bases in their 'pile of pennies' structure for DNA was almost identical to the spacing between individual amino acids in the polypeptide chains of the keratin proteins in wool. As a result, they were convinced that this near-identical spacing must have some profound significance for the interaction between proteins and nucleic acids, perhaps by allowing the two types of chain molecule to line up alongside each other during replication.

Figure by K. T. Hall. 'Pile of Pennies' reproduced with the kind permission of Cold Spring Harbor Laboratory Press.

thymonucleate. The result was a fibrous mass that could be drawn out of solution using a glass rod, dried, and mounted for analysis by X-ray diffraction.[36] The resulting X-ray diffraction patterns were almost identical to those seen with thymonucleic acid alone. Astbury interpreted the pictures as being positive evidence that Bell had successfully generated a nucleoprotein complex like those found in chromosomes and viruses.

Rosalind Franklin may have shunned making outrageous leaps of the imagination when interpreting structural X-ray data, but Astbury had no such qualms.[37] When interpreting the X-ray photographs of keratin in wool, his imagination had been an asset—it had enabled him to envisage how the keratin chain changed shape from an alpha to a beta form. Now, from one single piece of data—the measurement of the spacing between the bases in DNA—he had made another imaginative leap—one he was convinced would answer one of biology's biggest secrets: how living systems replicate themselves. It was the one instance when he would probably have most benefited from Bell's skill as his vox diabolica; and as irony would have it, this was the one time when this skill deserted her.

We now know that Bell and Astbury's model was hopelessly wrong. Nevertheless, that does not render it without worth. Measuring the distance between bases may have fired up Astbury's imagination into overdrive and led him to an erroneous model of DNA, but it certainly proved invaluable for someone else. Thirty years after Astbury and Bell proposed their model, James Watson acknowledged that it was Bell's X-ray photographs of DNA and her measurement of the 3.4 Angstrom spacing between adjacent bases that had provided him and Crick with a crucial foothold when they first began to consider building a model of the structure of DNA.[38]

The nucleoprotein model is also worth reflecting on, as it was indicative of a growing trend in biology. This was a movement away from the geneticist's idea of the gene as an abstract unit towards the notion that the gene might actually have a material existence. Work by cytologists using chemical staining techniques, combined with research into the genetics of the fruit fly, suggested that the determinants of genetic traits were arranged in a linear sequence along the length of the chromosomes. Now Astbury believed that his model could explain the linear sequence of genes that the cytologists were proposing ran the length of a chromosome.[39]

Once again, Astbury felt his overarching idea that the macroscopic properties of biological systems could be explained in terms of their underlying molecular structure had proved successful. From his initial work with Bragg on compounds such as tartaric acid, and then his work at Leeds on wool fibres, this conviction had come to define all his work. Now he took this idea to new heights with a molecular description of the nucleoproteins found in chromosomes and viruses. Convinced that this model offered a key insight into the most basic processes of living systems, he presented it at a 1938 meeting held at Cold Spring Harbor on Long Island as well as publishing the work in the journal *Nature*.[40,41]

In the same year, he was invited to attend a meeting of physicists and geneticists at Klampenborg near Copenhagen. The aim of the conference was that physicists and geneticists should come together in an attempt to address the question of genes. It was a theme that was continued in late August the following year, when Astbury was invited to speak at the 7th International Congress on Genetics held in Edinburgh. One of the organisers of the meeting, the geneticist Herman Müller, declared that its aim was to bring chemists and physicists together with biologists in order to 'answer questions that lie beyond the limits of biology', the major one of these being—what is a gene?[42] While geneticists such as Müller had focused on observing patterns of inheritance for traits such as eye colour in the fruit fly, it was felt that the techniques of chemistry and physics might be able to shed new light from other angles on the molecular nature of heredity. With his publication of Bell's X-ray photographs, Astbury was at the forefront of this new approach, which transcended boundaries between the traditional scientific disciplines.

The aim of the meeting was undoubtedly optimistic—to bring scientists from a diverse range of disciplines together with the hope that, by so doing, a fundamental question in biology might be solved. One report described the meeting as being 'a harmonious whole, moved by a common purpose, to serve mankind, to cultivate a fellowship'.[43] Sadly, the reality of the conference fell far short of its noble aims. Claiming that travel to Edinburgh would be impossible due to complications with visas, the Soviet delegation announced that they would be unable to attend.[44,45] The truth behind their absence is probably rather more sinister. The idea that inheritable biological traits might be determined by material entities and not conditioned by the environment was irreconcilable with the Marxist–Leninist ideology at the heart of the Soviet State. To keep in step with Marxist political doctrine, Soviet biology maintained that biological traits could be altered, or even improved, by controlling the surrounding environment. The chief proponent of this doctrine was Trofim Lysenko a biologist who, having won favour with Stalin, ensured that his own ideas about heredity became the established scientific orthodoxy in the Soviet Union. It was to leave a tragic legacy at both a national and individual level. Far from bringing about a workers' paradise, the implementation of Lysenkoist genetics in agriculture resulted in famines that left millions dead. Meanwhile, the ruthless and often-brutal oppression of any biologist who threatened this view forced many life scientists, including

Müller (who had organised the Edinburgh meeting), to flee Russia. With this knowledge, the Soviet excuse about visa complications becomes understandable. Attendance at the meeting would undoubtedly be interpreted as opposition to Lysenko's doctrine on genetics and thus would have come at a high price for any Soviet scientist who made the journey to Edinburgh.

While the scientists in Russia faced persecution, there were threats to the harmony and collegiate spirit of the scientific community far closer to home. As a result of the political situation in Germany, the mood in the wider world was becoming decidedly grimmer and more pessimistic. The same report on the meeting went on to say that 'murmurings outside grew in intensity, and towards the end of the second day the air became disturbed'.[46]

Despite its noble ideals, it was inevitable that the meeting would fall victim to the worsening international situation. The 34-strong German delegation received notification that it would be best if they returned home. At first, they were reluctant to do so, until it became apparent that if they remained in Edinburgh, they might face severe difficulties in getting home. At the same time, the delegation from the Netherlands were forced to return home as their army was being mobilised. The 130 scientists who had made the journey from the USA faced the prospect of a potentially dangerous journey home if war was declared or else being left stranded in Britain. At this point, the meeting became more than a scientific conference—with hourly enquiries being made by the local inhabitants about which nationalities were still in attendance at the meeting, it was now described as a 'barometer' from which inferences about what was happening on the world political stage could be reliably made.[47]

On 3 September, the British Prime Minister Neville Chamberlain made a radio address to the nation to deliver the grim news that, as he had received no reply from the German ambassador regarding Britain's demand that German forces withdraw from Poland, a state of war now existed between the two countries. As Europe was plunged into war for the second time that century, Astbury's attention was, somewhat understandably, diverted from DNA. The onset of war was accommodated by adopting a certain style of black humour. On hearing that a German bomb had fallen on his alma mater of Cambridge, he was outraged, declaring that 'it is one thing capturing Paris but it is quite a different thing defiling Cambridge'.[48] But such humour was all very

much bravado and bluster in the interests of putting on a brave face, for
in a letter written to the vice-chancellor in the same year he conveyed
his very real fears:

> I rather think that many of us ought to get, and would like to have, some
> practice in the use of defensive weapons, in case things should get desper-
> ate, as seems not unlikely at the moment, and I should feel happier in my
> mind if the University could organise something of the sort on an official
> basis.[49]

Inevitably, the war brought disruption to the laboratory. Most of his
staff were required to do War Service and Astbury himself served first
with the Home Guard before going on to teach air navigation as a Pilot
Officer in the RAF Volunteer Reserve.[50] Maintaining an active research
group proved difficult under such conditions. In a letter written to
Isidore Fankuchen, the X-ray crystallographer with whom Bernal had
done the work on TMV at Cambridge, Astbury's sense of exasperation
at the interruption to his work was all too apparent:

> there are far too many damned things to do, that's all, and I have no
> secretary, no personal assistant and not even a lab-boy now! What a war![51]

A major blow came when Florence Bell was summoned for military
service. In desperation not to lose her, Astbury wrote to the War Office
explaining that her skills were indispensable and begging that she be allowed to remain in his laboratory.[52] But his letter was in vain and in
1941, Bell left the laboratory to serve as a radio officer in the Women's
Auxiliary Air Force (WAAF). Astbury wrote in desperation to secure
funding for her salary from the Rockefeller Foundation so that 'we shall
at least have something to start afresh with when the war is over', and
in recognition of the important work that Bell had done in Astbury's
laboratory, the University of Leeds extended her post should she wish
to return after her war service.[53,54] But by this time, other more powerful forces were shaping Bell's life. In January 1943, she wrote to A. E.
Wheeler, the registrar at the University of Leeds, to thank him for keeping her post open but informing him that on 21 December 1942, she had
married Lieutenant J. H. Sawyer of the US Army and now intended to
join her new husband in the USA.[55] Bell left Britain to begin a new life
in the USA, where she worked first for the British Air Commission in
Washington, DC and then later as an industrial chemist for the Magnolia Petroleum Company in Beaumont, Texas.[56] Her departure was

certainly a blow to Astbury's fortunes and stalled the momentum of his work on nucleic acids. But by the time that the War finally ended, he did indeed 'have something to start afresh with'—not new financial support or equipment, but something possibly even more valuable.

On 6 June 1944, Allied troops poured ashore onto the beaches of Normandy to begin the liberation of Europe from the Nazis. The Nobel Laureate Joshua Lederberg later said that the D-Day landings were one of two key events that year that shaped the course of history.[57] The other was a scientific paper published by a quiet, shy clinician called Oswald Avery. Working at the Rockefeller Institute in New York City, Avery's work transformed our understanding of nucleic acids, and the impact of this work was later recognised to be so great that one popular undergraduate textbook hailed it as 'Avery's Bombshell'.[58] Astbury praised this work as being 'one of the most remarkable discoveries of our time' and, following the gloom, hardship, and fear of the war years, it could not have come at a more opportune time.[59]

7

'Avery's Bombshell'

To hear his work described as a 'bombshell' and placed on a par with the D-Day landings in terms of its impact on history would probably have made Oswald Avery raise his eyebrows (Figure 16). His manner was always one of modesty and restraint, both of which were reflected in his physical appearance, as captured eloquently in the following recollection by his colleague Rene Dubos:

> Everything about his person was in a low key that made him inconspic-
> uous, like the buildings in which he worked and lived. He was small and
> slender and probably never weighed more than 100 pounds. In behaviour
> he was low-voiced, mild-mannered, and seemingly shy. His shirts, suits,
> neckties and shoes were always impeccable but were never as subdued as
> his physical person.[1]

This modest and restrained demeanour was also reflected in the way that Avery approached science. In stark contrast to Astbury, who cast his net wide over a diverse range of experimental subjects and had no reservations about making grand inferences from very limited data, Avery's approach was to focus tightly on one very narrowly defined experimental question. In fact, he was so keen to avoid anything that seemed like an unnecessary distraction from the business of research that he even shunned scientific conferences—attendance at which is often considered essential for building the professional networks that promote career advancement—offering instead the excuses of lack of time and funds, or poor health. Dubos said that Avery:

> conducted his investigations with the least possible expenditure of phys-
> ical effort, and with strict economy of materials, laboratory equipment
> and experimental animals. For him, the ideal experiment was one that
> yielded a clear and inescapable conclusion from a limited number of facts
> observed in a few test tubes or a few animals.[2]

Figure 16 Oswald Avery at work in his laboratory.
Photograph by George Hirst. Courtesy of Rockefeller Archives Center.

While Astbury accumulated accolades, gave public lectures and courted industry, Avery avoided any engagement that he deemed to be unnecessary in an effort to devote all of his time and talents to his small group of researchers. Before starting a piece of work at the laboratory bench, Avery would often sit for days, quietly thinking through the problems and challenges ahead. Only when he was satisfied that he had contemplated the problem to the point of exhaustion would he rise from his desk and begin laboratory work, still maintaining a calm air of reflection as he began gathering test tubes and pipettes while gently whistling to himself the lonely tune of the shepherd's song from Wagner's opera *Tristan and Isolde*.[3] Avery's approach to how science should be done at the bench is perhaps best expressed in an address that he gave in 1949 in honour of his colleague and flatmate Michael Dochez, who had just been awarded the Kober Medal of the Association of US Physicians.

In his speech, Avery praised the way in which Dochez carried out his science, yet, on reading his words, one cannot help but feel that Avery was also quietly articulating his own personal philosophy:

> I have never seen his laboratory desk piled high with Petri dishes and bristling with test tubes like a forest wherein the trail ends and the searcher becomes lost in dense thickets of confused thoughts. I have never seen him so busy taking something out of one tube and putting it into another there was no time to think of why he was doing it or of what he was actually looking for. I have never known him to engage in purposeless rivalries or competitive research. But often I have seen him sit calmly by, lost in thought, while all around him others with great show of activity were flitting about like particles in Brownian motion; then, I have watched him rouse himself, smilingly saunter to his desk, assemble a few pipettes, borrow a few tubes of media, perhaps a jar of mice, and then do a simple experiment which answered the very question he had been thinking about when others thought he had been idling in aimless leisure.[4]

Avery's sanguine, almost Zen-like philosophy of the practice of science may seem quaint in an age where 'multitasking' is revered as an indispensable skill, and it would probably not go down well today with an overworked head of department eager to achieve a top score in a research assessment exercise, but it certainly proved a success for him in the 35 years that he worked at the Rockefeller Institute Hospital. The institute had been established to provide a research environment where scientists would be liberated from the burdens of administrative duties and teaching commitments, and, right up until his retirement in 1948, Avery enjoyed a successful career there.

The main focus of his research during his many years at the Rockefeller Institute was the study of the pneumococcus bacterium, which was the causative agent of lobar pneumonia. Avery had become interested in this bacterium when he had joined the Rockefeller Institute Hospital (later to become the Rockefeller Foundation) in 1913.[5] At that time, pneumonia was a leading cause of death in the USA, killing more people than diphtheria, scarlet fever, and typhoid fever combined, and a concerted research effort was focused on developing an antiserum against the bacterium.[6–8] During the First World War, the problem of pneumonia became even more important as there were frequent outbreaks of the disease in army camps, where the condition was often a secondary complication arising from infection by pathogens such as the influenza virus.

One of the most important developments in the study of the pneumococcus bacterium was thanks to the work of British microbiologist Fred Griffith. As a medical officer in the pathology laboratories of the British Ministry of Health, in 1923 Griffith discovered that, like many other bacteria, pneumococcus existed as two distinct strains that could be distinguished by their particular appearance when viewed under a microscope.[9,10] One strain had a smooth, glistening surface and was termed the 'S strain', while the other appeared dull and rough, and became known as the 'R strain'. It was a distinction that had important clinical consequences. The smooth appearance of the S strain was because each cell was surrounded by a capsule of polysaccharide molecules that allowed it to escape being devoured by the white blood cells of an infected host's immune system. As a result, only the S-strain cells were able to cause disease, while the R-strain cells, which lacked this polysaccharide capsule, were avirulent. Depending on the particular kind of polysaccharides that made up this capsule, the S strain could be designated as belonging to one of three particular types, numbered I, II, or III, each of which triggered the production of a different set of antibodies when they infected a host.

Like Avery, Griffith had a shy, retiring manner to the extent that when he presented a paper at an international meeting of microbiologists in 1936, his voice was so faint that no one in the auditorium could hear a word.[11] Yet the modesty of Griffith's manner did a disservice to the importance of his work. Griffith wondered what would happen if mice were injected with a mixture of live R-strain cells, which did not cause disease, and dead cells from the disease-causing S strain that had type II surface polysaccharides. What he found surprised him. Despite being injected with the avirulent R strain of the bacterium, the mice still died from pneumococcal infection. When the blood of the dead animals was analysed, it was found to contain live, virulent S-strain bacteria of type II. From this, Griffith concluded that somehow the dead type II S-strain bacteria had converted, or 'transformed', the live R-strain bacteria into becoming virulent.[12] How had this happened?

Griffith's career was tragically cut short when he was killed while 'fire-watching' during an air raid in 1941, but others continued his work.[13] In New York City, the clinicians Martin Dawson and Richard Sia showed that transformation did not even require the bacteria to be injected into a live animal—it could occur just as easily when live R-strain and dead S-strain bacteria were grown together on an agar dish.[14],[15]

One possible explanation for this effect was that the ability to synthesise capsular polysaccharide of a particular type and so become virulent was somehow being passed on from the dead S-strain cells to the live R-strain cells. Dawson and Sia were not clear as to how this could be happening, but their observations of the effects of environmental influences on transformation gave them a possible clue. Both the application of heat and the action of certain bacterial enzymes caused significant loss of transforming activity in the bacteria, suggesting that some material substance might be at the heart of the process.

If there was indeed some substance within the pneumococci that could confer virulence on bacteria of the R strain, then perhaps it could be extracted from bacterial cells and used to induce in vitro transformation as Dawson and Sia had observed. This task fell to Lionel Alloway, another clinician also based at the Rockefeller Institute. Alloway grew large cultures of type III S-strain pneumococci, and, after collecting them by centrifugation, he broke the cells open by rapidly freezing and thawing them to release their contents. When he passed these cellular extracts through a filter and added the filtered solution to a culture of R-strain bacteria, he found that, sure enough, the organisms became transformed into disease-causing S-strain pneumococci of type III. His conclusion was that some substance within the cellular filtrate had passed on the ability to synthesise type III polysaccharide and become virulent.[16]

Unfortunately, however, Alloway found that he could not always reliably reproduce these results. This was presumably because his method of making cellular extracts was quite crude and the resulting filtrates were often contaminated with other cellular material that reduced their ability to induce transformation. What was required was a way of purifying the transforming substance from the cellular extract and then identifying its chemical nature.

It was to this task that Avery turned his attention. He began by growing S-strain pneumococcus of type III on a large scale in cultures of 50–75 litres. After 16–18 hours' growth, the bacteria were collected by spinning the cultures in a centrifuge, which was chilled to minimise degradation of the cellular contents. The resulting pellet of bacteria was resuspended in chilled salt solution to give a thick creamy suspension that was then heated to 65 degrees Celsius to deactivate any intracellular enzymes that might destroy the transforming principle. After several washes in salt solution, the cells were incubated with sodium deoxycholate, a detergent that burst open the cells and released their contents into the solution. Proteins were removed by the addition of chloroform and polysaccharides were digested away using bacterial enzymes. The remaining solution was then treated repeatedly with chilled alcohol, which resulted in the precipitation of white fibrous strands that could be wound around a stirring rod.

When added to a culture of R-strain pneumococcus, these purified fibres resulted in transformation of the organisms into type III S strain. The result was also highly reproducible. But what exactly were these fibres? When subjected to specific chemical tests that detected protein, the fibres gave a negative result and, when heated to 65 degrees Celsius, a temperature at which proteins lose activity, the fibres retained their transforming properties. Treatment with the digestive enzymes trypsin and chymotrypsin, which break down proteins in the gut, similarly had no effect on the ability of the material to induce transformation. On the basis of these results, it was starting to look as if Avery could rule out the possibility that the fibres were protein.

If it was not a protein, then was the fibrous material a nucleic acid? Chemical analysis showed that the fibres were composed of carbon, nitrogen, oxygen, and phosphorus in almost exactly the same proportions that had been calculated theoretically for deoxyribonucleic acid. When treated with enzymes that were known to degrade DNA, the material lost all its transforming activity. Meanwhile, enzymes that degraded RNA, the one other known type of nucleic acid, had no effect.

When the material was spun in analytical ultracentrifuge, it gave a tight band, suggesting that it was composed of molecules of a defined length and its molecular weight was estimated to be 500,000—of the same order as thymonucleic acid. The final piece of evidence came from studying how the fibres absorbed ultraviolet (UV) light. Depending on their respective chemical composition, different kinds of molecules absorb UV light maximally at different wavelengths. This means that the particular absorption pattern of UV light gives a kind of fingerprint for a particular type of molecule. When the fibres of Avery's transforming substance were analysed in this way, he found that they absorbed UV light at exactly the same wavelength as nucleic acids. The combination of results from these different experiments led Avery to conclude that, far from being merely structural components of the cell made from a monotonous repeat of tetranucleotides, nucleic acids were 'functionally active substances in determining the biochemical activities and specific characteristics of pneumococcal cells'.[17]

Avery's results suggested that, in pneumococcus at least, nucleic acids alone and independent of any protein were capable of passing on a specific biological trait that transformed bacteria from one type into another. But he was reluctant to extend this conclusion any further. He was a man characterised by caution in all aspects of his life, a good example of which was given by his colleague Dr Yale Kneeland of the Columbia-Presbyterian Medical Center. Kneeland liked to relate a story about how he and Avery were driving together from Manhattan to Long Island. Sitting in the passenger seat, Avery glanced down at the dashboard and, seeing the dial showing 80, asked nervously 'Don't you think that we're travelling a bit too fast, Dr Kneeland?' Glancing down, Kneeland reassured Avery that he could relax, pointing out that he had confused the radio dial for the speedometer, which was actually showing less than 40 mph.[18]

This caution extended into the interpretation of his scientific results, and he was loathe to make grand claims for his work. As he used to advise his students: 'It is lots of fun to blow bubbles, but it is wiser to prick them yourself before someone else tries to'.[19] It was advice which made a sharp contrast with the giddy way in which Astbury and Bell had seized upon the correspondence between the near-identical spacing between bases in DNA and amino acids in keratin and insisted that it must have some grand significance for how proteins and nucleic acids function together in replication.

Despite Avery's modesty and restraint at the time, his experiment is now hailed in biology textbooks with dramatic terms such as 'bombshell' and is told as the story of a discovery that, despite being revolutionary, was ahead of its time and so went neglected and ignored.[20] When told in this way, the real mystery seems not to be how the nucleic acids caused the transforming effect but rather that no other scientists seemed to take any notice of Avery's work at the time.[21] It has been suggested that, since Avery published his results in a medical journal that was read only by clinicians, his work remained unknown to biochemists and microbiologists.[22] In such an account, Avery becomes rendered into a classic cliché in the history of science—that of the lone genius—the scientist who is somehow 'ahead of their time', whose work, despite being of phenomenal importance, arrives on the scene prematurely and as a result is ignored.[23]

A closer study, however, reveals that Avery's case was rather more complex. The account often given in textbooks in which science proceeds through moments of revolution is an appealing one, if only for its sense of implicit drama rather than any historical truth. At first sight, Avery's identification of nucleic acid as the carrier of biological traits in pneumococcus seems like a prime candidate for such a moment of revolution. But more often than not, revolutions are constructed with the benefit of hindsight, and they require two criteria: firstly that a pre-existing order is overthrown, and secondly that this occurs in a very short space of time.

Avery's work did neither. There was no overarching dominant order that was brought crashing down by his discovery. True, many scientists believed, for good reasons, that proteins were the carrier of heredity material, but others such as Astbury already acknowledged the possibility that nucleic acids might have an important role in replication. Secondly, Avery's results did not signal the immediate and overnight dismissal of proteins in favour of nucleic acids as the genetic molecule. He faced strong criticism from other scientists like his fellow Rockefeller colleague Alfred Mirsky, who felt that the evidence still favoured proteins as the active molecular agent in heredity. Even in the early 1950s, Avery and his colleagues were having to defend themselves against criticism from Mirsky, who argued that, since it was impossible to be absolutely sure that Avery had removed every trace of protein from his nucleic acid samples, the transference of traits in pneumococcus might still be due to residual protein and not nucleic acid.[24] Avery himself even admitted this possibility, saying in an annual report written in 1946–47 to the Board of Scientific Directors:

> From the beginning we ourselves have been keenly alert to the possibility that the presence of some substance other than the deoxyribonucleate in our preparations may be responsible for the biological activity.[25]

Avery's conclusions gained considerable support in 1952 from two researchers working in the field of bacterial viruses, or bacteriophages. Alfred Hershey and Martha Chase, by labelling viral nucleic acids and proteins with radioactive phosphorus and radioactive sulphur, respectively, showed that it was only the nucleic acid portion of a virus that entered a host cell on infection, while the viral protein remained outside.[26] The interpretation was that it must be due to the nucleic acid alone that the virus was able to replicate itself on entering the host cell; proteins played no role.

Hershey and Chase's work is often cited as having confirmed Avery's earlier work and providing the evidence that finally silenced his critics. But it was certainly not the case that Avery's work had languished in obscurity until 1952 when Hershey and Chase's work finally gave it credibility. Far from being doomed to obscurity in a clinical journal that no one bothered to read, Avery's paper had certainly attracted attention in the intervening years. It was the subject of discussion at a number of scientific meetings held in the USA during 1945–1947, such as the Cold Spring Harbor Symposium of 1946 on 'Heredity and Variation in Micro-organisms', which, despite his reclusive nature, Avery had actually bothered to attend.[27] Within only a few years, microbiologists and clinicians in both France and the USA were attempting to replicate Avery's findings in other bacteria, for example, *Escherichia coli* and *Haemophilus influenzae*.[28–34]

Feeling that his best scientific years were behind him, Avery retired to Nashville in 1948 and did no more work on nucleic acids. As was his manner, his concern had always been to focus on one very specific question—what the chemical nature was of the transforming principle—and not be distracted by the questions that might follow. As he said in a 1943 letter to his brother Roy:

> with mechanisms I am not now concerned. One step at a time and the first step is, what is the chemical nature of the transforming principle? Someone else can work out the rest. Of course the problem bristles with implications.[35]

Others did indeed eventually 'work out the rest'. In his memoir, James Watson acknowledged that it was Avery's paper of 1944 on pneumococcus that inspired him to focus on DNA as the material agent of heredity.[36] Avery's paper had raised two fundamental questions. First, was the ability of nucleic acids to cause transformation confined only to pneumococcus? And second, what was the mechanism by which it achieved this effect? While microbiologists in the USA and France tackled the first question, at least two scientists in Great Britain had turned their attention to the latter. One of these was the Nottingham-based chemist J. M. Gulland, who said that Avery's work marked 'the first occasion on which specific transformation has been experimentally induced in vitro by a chemically defined substance, and its implications are of the greatest importance in the fields of genetics, virology and cancer research'.[37]

The other was Astbury, who had once met with Avery briefly while on the same lecture tour of the USA in 1937 where he first met Pauling. Through his work with Preston on the alga *Valonia*, he had become interested in the polysaccharide chains found on the surface of certain cells and was therefore keen to discuss the surface polysaccharides of the pneumococcus bacterium with Avery.[38]

Astbury had cultivated his interest in the pneumococcus through a collaboration with the biochemist W. T. J. Morgan, who was working at the Lister Institute of Preventive Medicine in London.[39–41] In 1944, Morgan published a paper in *Nature* that summarised Avery's findings, describing them as an example of 'controlled mutation'.[42] A year later, Astbury described Avery's work using exactly the same phrase, suggesting that he had read Morgan's paper and that it was most likely this article that had made him aware of the work at the Rockefeller.[43]

But whereas Avery trod with caution regarding the implications of his own result, Astbury could barely contain his excitement:

> In this connection I was terribly thrilled to read of Avery's identification of the factor that predisposes the protein core of the pneumococcus to build its specific polysaccharide capsule (and ever after its own supply of factor too!). This is one of the most remarkable discoveries of our time and that the factor should turn out to be a bare thymonucleate after all touches me very closely for just before the war we did quite a lot of X-ray work on sodium thymonucleate and the nucleic acids in general.[44]

Before discovering Avery's work, Astbury had been confident that biological replication was controlled by a hybrid structure composed of both protein and nucleic acid, or a 'nucleoprotein'. Others had developed Astbury's model even further.[45] The biochemist Kurt Stern argued that, in the kind of nucleoprotein hybrid that Astbury had proposed, it might well be the nucleic acids that were the more important of the two components. In Stern's model, the different bases running along Astbury's single chain of nucleic acid could well satisfy the two criteria posited by Schrödinger and generate enough structural variation to specify particular biological traits. In what seemed to be a tantalising moment of prescience, Stern even suggested that the shape of this single nucleic acid chain might be twisted into a spiral. The problem would be, however, that as the bases were free to swivel around the axis of the nucleic acid chain, any pattern of variation in base sequence would be unstable and rapidly lost as the bases spun around. This was where the protein chain came in. The polypeptide chain, claimed Stern, acted to lock the bases on the nucleic acid chain into fixed positions. In this way, Astbury's hybrid nucleoprotein model formed from a single chain of nucleic acid and polypeptide could easily carry biological information.

Avery's result presented a formidable challenge to this model, for he had shown that (in pneumococcus at least) nucleic acid alone was quite capable of carrying information without any need for an accompanying protein chain to lock the bases in position. Somehow, the ability to do this must reside in the structure of the nucleic acid itself, and Astbury believed that he had glimpsed a way in which this might be possible.

The answer, he thought, lay in the polysaccharide chains that coated the surface of the pneumococcus bacterium. Astbury had wondered whether these surface polysaccharides might be the reason why pneumococcus could be classified into different types such as I, II, and III. Writing to Assistant Director of the Natural Sciences Division at the Rockefeller Institute F. B. Hanson, Astbury elaborated on his idea:

> A very exciting thing (to me, at least) happened the other day. I saw in a book an electron microscope photo of a pneumococcus, and ... I conceived the idea that the specificity of the polysaccharide capsule of the pneumococcus might rest not on chemistry alone but also on the geometrical arrangement of the chain-molecules constituting the capsule.[46]

Astbury was suggesting that the different types of S-strain pneumococcus might arise not just because their respective surface polysaccharide macromolecules had a differing chemical composition, but also because they had different *shapes*. Once again, he was proposing that the characteristics and function of living systems could be reduced to one key feature of their biological macromolecules—namely, their precise three-dimensional shape.

And what was true for the surface polysaccharides of pneumococcus might also be true for its nucleic acid. Perhaps it was the particular way in which the nucleic acid chain was folded that enabled it to carry specific biological traits? If so, then these differences in molecular geometry should be readily detectable using X-ray diffraction. It was a tantalising possibility and an insight that Astbury felt brought him to the verge of grasping one of biology's biggest secrets.

Eager to put his ideas to the test, Astbury wrote to Avery requesting a sample of the purified thymonucleic acid for analysis by X-ray diffraction:

Dear Professor Avery

I do not know whether you will remember me, but I had the pleasure of having a little talk with you about the pneumococcus and things when I was lecturing in the States in 1937. I have recently been extremely thrilled by your identification of the factor that can transform the rough variant of the pneumococcus into the smooth specific form of another type and I am writing to ask if you could possibly let me have some for X-ray examination. You may know that I have done a fair amount of X-ray work on the structure of the nucleic acids, though on account of the war, some of it is unfortunately not yet published, and in particular I have been able to obtain quite a lot of information from my photographs of highly polymerised sodium thymonucleate. I note that you identify your factor as being probably this kind of thing, and I think you will agree that it has now become a matter of considerable urgency to get down much more thoroughly to the question of the specificity of the nucleic acids. If I could get some decent X-ray photographs of your preparation too, it might turn out helpful and I should be very grateful indeed if you could supply the stuff . . . I do hope you can let me have this stuff, with all relevant instructions. It seems to me that there is a wonderful chance here to make an important step forward.[47]

Whether Avery ever sent the requested material or even replied to Astbury's letter is highly unlikely. In keeping with his shy, reclusive manner, Avery rarely responded to correspondence and carefully destroyed many of his files and papers when he retired to Nashville.[48] Nevertheless, Avery's results were such an inspiration to Astbury that they left him wishing desperately for:

a thousand hands and labs with which to get down to the problem of proteins and nucleic acids. Jointly those hold the physico-chemical secret of life, and quite apart from the war, we are living in a heroic age, if only more people could see it.[49]

As the Second World War came to an end, there came a golden opportunity to turn this from a mere wish into a concrete reality.

8

'Nunc Dimittis'

The end of the Second World War brought a mood of welcome relief and optimism. As the formidable task of rebuilding Europe from ruin began, there was a sense among politicians that this was the opportunity for a fresh start, a chance to build a better society. In Great Britain, this mood manifested itself in the Beveridge Report, which proposed the creation of a 'cradle-to-grave' system of welfare, including free secondary education and a free national health service.

But this optimistic spirit was by no means confined just to the political world. In science too, there was a sense that, with the end of the war, the time had come for research projects that were grand in scale and ambition. For Astbury, there was no grander task than to bring the methods of physics and chemistry to bear on biology and, in particular, the most fundamental question of the life sciences—how do biological structures replicate themselves? Avery's results had fired his imagination and left him convinced that nucleic acids, as well as proteins, were somehow at the heart of this process and that it was by the methods of physics that the mystery would ultimately be solved. In 1945, he wrote to the vice-chancellor at Leeds proposing that Preston, his old colleague in the Department of Botany and with whom he had worked on the structure of *Valonia*, be appointed to a professorial chair. But, as well as offering support for Preston, Astbury used the letter to make the point that the future of the life sciences, as he saw it, was in molecular biology:

> Preston is a pioneer of the new outlook in botany. I mean that botany, as indeed all biology, is now passing over into the molecular structural phase . . . In all branches of biology and all universities this thing must come to pass and I suggest that Leeds should be bold and help to lead the way.[1]

Astbury's view that biology must embrace a molecular approach, rooted in the methods of physics, was vindicated when he was invited by the Council of the Royal Society in 1945 to deliver the prestigious

Croonian Lecture. Dating from the very first days of the Royal Society in 1684, this was an annual lecture normally given by a biologist on the theme of the life sciences. That a physicist had, for the first time, been invited to give this address was a powerful testimony to the impact that Astbury's work was making on the life sciences, and it was an honour that he was delighted to accept. Delivering the Croonian Lecture would, he said, present an important opportunity to bring about the 'impending final breakdown of the artificial barriers between biology and the basic sciences'.[2]

Moreover, he was determined that his own laboratory should lead the way in breaking down these barriers. With himself at the helm, Leeds would host the country's first department of molecular biology and become established as the national centre for this new discipline. This new department would not only blaze the trail across this exciting new frontier but also serve as a personal memorial to his life's work and legacy, a sentiment that he described rather more poetically in a letter to Hanson at the Rockefeller Foundation:

> It is my great dream (and the Foundation's, I feel sure) to found something of the sort before I finally pass out. I am labelled a physicist, but all my efforts have long been devoted to the welding together of the sciences, and in particular to the union of biology with physics and chemistry; and if only I can persuade the people here to strive whole-heartedly towards the same end, then I can say my 'Nunc dimittis'.[3]

The 'Nunc Dimittis' is a prayer that is sung at evening services in the Roman Catholic and Anglican traditions and sometimes at funeral services. Based on a passage from the Gospel of St Luke, it begins with the words 'Lord, now lettest thou thy servant depart in peace' and beautifully expresses the sense of peace derived from the knowledge that one is departing having seen certain hopes and aims fulfilled.[4] In a more secular age, the same sentiment might be conveyed with talk of 'having achieved closure', though the latter expression has neither the same aesthetic nor the imaginative appeal of the former.

In May 1945, as the nation celebrated VE Day, Astbury had another reason to feel jubilant. That same month, the University of Leeds Senate had passed a resolution proposed by Professor Speakman of the Textiles Department to appoint a committee for the establishment of Astbury's new department, and, at their next meeting, it was recommended that Astbury be given a professorial chair as its new head.[5,6]

Astbury's new title would be 'Professor of Biomolecular Structure' and as a career move this should have delighted him, but there was one setback. Nowhere in the title of his new department was there any mention of 'molecular biology'.

Henry Kissinger is said to have once observed that, if academic politics are sometimes vicious, this is only because so very little is actually at stake—and the disputed title of Astbury's new department might seem to bear this out. Nevertheless, although it may seem at first to be nothing more than academic pedantry, it does give an interesting insight into how certain biologists felt towards (what were perceived to be) outsiders from other scientific disciplines encroaching on their field. Astbury might well have hailed his colleague Preston as a pioneer of the new outlook in botany, but there were others who gently reminded him that, as a physicist he was too 'insufficiently staffed on the biological side' to justify being head of a department with 'biology' in the title.[7] Despite the honour of being invited to give the Croonian Lecture, Astbury found himself having to defend the name 'molecular biology' against certain critics from within biology right up until his death in 1961.[8–10] Perhaps there was a feeling on the part of some biologists that their subject was being hijacked or usurped by outsiders with no qualification or experience of the life sciences.

Not willing to let such splitting of hairs over a name derail his grand plans, Astbury conceded to being unable to use the name 'molecular biology' in the title of his new department—but only begrudgingly so, as he made quite plain in a letter to Weaver at the Rockefeller Institute:

> I must confess that I am rather proud that you are in the process of adopting the phrase 'molecular biology', which I believed I was first to coin. And I am still sad that I could not get the people at Leeds to accept this name for my new department. Of course, you know how it is at a University: every member of the Senate had some queer objection to my suggestion (and naturally the biologists would not grant that I was in any way a biologist) and the result was that rather ridiculous mouthful, 'Biomolecular Structure'. But how simple and expressive of everything that we want to do is the name 'molecular biology'.[11]

The semantics of the new department was the least of Astbury's troubles. A far bigger problem was how to fund it. In the austerity of post-war Britain, money was not likely to be forthcoming for such a grand project. For the past twelve years, Astbury's work had been mostly supported by the Rockefeller Foundation with a grant of £2000 per annum, but with his current grant due to expire in the next year, he now found himself desperately looking around to secure new sources of funding. Astbury was grateful to the Rockefeller Foundation—they had supported him throughout his early years and had allowed him to reach a point where he had proved himself. He had reservations, however, about being supported for so long by money from abroad and felt that it was better to seek long-term funding from an institution closer to home.

A golden opportunity to secure long-term funding arose late in 1945 when Astbury found himself at a Royal Society dinner sitting next to Sir Edward Mellanby, chairman of the Medical Research Council (MRC). Aware of Astbury's international standing, Mellanby was interested in his work and wanted to hear more about his vision, which Astbury eagerly articulated in a letter to him:

> Believe me, Sir Edward, for years now I have had this dream of bring-ing physics and chemistry into closer relation to the needs of biology, and professor or no professor, I do want to establish this new venture on durable lines—call it 'biomolecular structure' or what you will, but you know well what I mean. The time is ripe, and every day it becomes clearer that there is a real hunger among biologists for fundamental molecular knowledge. I want to help to meet this demand, and to break down the barriers once and for all; and I ask you to help me in bringing this about. This is the first time that such a Chair as Biomolecular Structure has been instituted in our own country, and I do so want to make it a real thing and the forerunner of others of the kind.[12]

Mellanby felt that Astbury's work was a definitely an area the MRC should support, and he invited Astbury down to London to present his case to the MRC Funding Committee on 15 February 1946. A month later, Astbury received a reply from Mellanby regarding the decision. The news was not good:

> I think it is only fair to tell you that we are floundering in rather deep wa-ter about the proposals for giving you financial aid and I think you would be well advised to carry on with your own plans for obtaining money and assume that we should be able to give you little or no help, at least for the time being. Needless to say, I am very much disappointed at the situation, but it is no good for you or me to live in a fool's paradise.[13]

In his reply, Astbury found it difficult to hide his disappointment:

> Your letter, I must confess, was a pretty awful blow and a shock to my
> confidence that will take a bit of getting over. I am not blaming you
> of course—that would be utterly ungracious—for it is obvious that if
> anyone is on my side, you are. It is hard though not to feel a bit dis-
> tressed that one should be acclaimed as one of the 'Fathers' of textile
> science in this country and not to receive financial support from indus-
> try, and also to have contributed so much to the very fundamentals of
> medical science and yet to be turned down there too. However, there
> it is, I suppose, and the sooner I set about the cap-in-hand business the
> better. [. . .] Please believe that I am terribly grateful for you efforts in my
> behalf. And please don't feel that I shall wilt permanently under the blow.
> There are so many wonderful things still waiting to be done with the bi-
> ological molecules, and by hook or by crook, I'm determined to do some
> of them.[14]

The reasons why the MRC chose not to support Astbury are not entirely clear. One of the only sources relating to the incident is a letter
written many years later by Sir Harold Himsworth, secretary of the
MRC from 1949 to 1968.[15] In his letter, Himsworth speculates that, during his meeting with the MRC, Astbury placed too much emphasis on
the study of synthetic macromolecules such as those found in plastics,
as if he were suggesting that these simpler chain molecules should be
studied in depth before attempting work on the more complex biological macromolecules. It is certainly true that a significant portion of his
research was involved with synthetic chain molecules in collaboration
with companies such as British Celanese and ICI, but, considering how
excited he was when he learned of Avery's work, it is difficult to believe that he would have neglected to mention the central importance
of proteins and nucleic acids. Astbury himself seemed to think that his
work had not been perceived to be sufficiently medical, for, as he explained to Dr H. M. Miller of the Rockefeller Foundation a few months
later:

> I am rather between the devil and the deep sea—biology and industry (I
> will leave it to you to argue out which is which)—though I had hoped
> ultimately to find support definitely from medical funds, e.g., the Med-
> ical Research Council. However, it appears now that this is not going to
> work—I am too fundamental, or molecular, or something, to justify the
> expenditure of medical money on me—and I am now driven to scout
> around desperately once more for help, and something more permanent

this time, too, so that I can at least have a few years' peaceful research free from these exhausting financial worries.[16]

Perhaps part of the problem was that, by now, other players had entered the field of molecular biology and were establishing their own centres of research. One of them was Bernal, Astbury's old colleague from the Royal Institution and whose biophysics unit at Birkbeck College, London was funded by the Nuffield Foundation and was opened by Lawrence Bragg in 1948. But of far greater significance to Astbury's work was the physicist John Turton Randall, whose career path had reflected Astbury's own in several ways. Like Astbury, Randall had humble origins in the north of England. The son of a Lancastrian market gardener, he had worked his way from a local grammar school to the University of Manchester.[17] As with Astbury, the Braggs had also proven to be a powerful influence in shaping the future course of Randall's career. While at Manchester, Randall came into contact with Lawrence Bragg, who, despite the fact that Randall showed enormous promise as a scientist, encouraged him to get a job in industry rather than stay on in academia to undertake research for a doctorate. It was a move that paid off, as Randall quickly became Head of Research at the General Electric Company (GEC). In 1937, however, he returned to academia, taking a fellowship at the University of Birmingham, where two years later he began working on a problem that, it might well be argued, was to be the salvation of the nation.

This was the problem of how to boost the strength of radar. Together with his fellow physicists H. A. H. Boot and J. Sayre, Randall developed a device known as the cavity magnetron, which enabled the production of an intense beam of low-frequency electromagnetic radiation. The device was central to the radar systems used during the Second World War. As it was very light, it could easily be carried aboard both fighter and bomber aircraft and was essential during missions flown in poor weather and at night. It was also deployed on anti-aircraft guns and in the detection of German submarines in the battle of the North Atlantic. In fact, President Franklin Roosevelt hailed the cavity magnetron as 'the most valuable cargo ever to reach these shores'.[18] It also proved to be highly valuable to Randall. Having in put in a claim to the Royal Commission on Rewards for Inventors (a body that awarded money for inventions deemed to have served the national good) in 1949, he received one-third of the total £36,000, which was a sum on which he had also secured tax relief and compensation for legal expenses incurred.

This made him wealthy beyond the dreams of most academics and provided a degree of independence that gave him a certain freedom in his research. After leaving Birmingham, Randall was appointed Head of Physics at St Andrews University in Scotland, where, like Astbury and Bernal, he began to develop an interest in applying the methods of physics to the study of living systems. Randall soon began building up the Physics Department at St Andrews, but a far better opportunity soon presented itself. Allan Nunn May, the most likely internal candidate to be appointed Head of Physics at King's College, London, had just been arrested. During the war, May had been in contact with atomic scientists and had been found passing secret information to the Soviets. May protested that, as the Russians had been fighting with Allies against the Nazis, he could hardly be charged with spying—an argument that failed to convince the authorities. As a result, May was imprisoned for ten years and Randall was appointed to the post, where he set about establishing his Biophysics Unit on the site of a bomb crater in the college quadrangle.[19]

While applying for research funding in his early days at St Andrews, Randall had once been told that he had asked for too little money. It was a mistake that he was determined never to repeat.[20] In stark contrast to Astbury's fortunes, when Randall approached the MRC with a bold request for funding, he was awarded the princely sum of £22,000 to establish his Biophysics Unit at King's.

In a letter to Randall, Astbury said that he was pleased to hear that things were going well for Randall's new department at King's, but his complaint in the same letter that 'my own research life has been one long struggle for years' was very telling.[21] Writing many years later, Mansell Davies, a member of Astbury's own research team at the time recalled that behind this public display of magnanimity, Astbury hid private feelings of great disappointment. Davies had joined Astbury's laboratory in September 1942 as a researcher working on artificial fibres for the company British Celanese, before taking up an ICI Research Fellowship to work with Astbury on using the new technique of infrared spectroscopy to study protein structure. Davies became a valued member of the team and Astbury went on to support him throughout the rest of his career. When Davies applied for a lectureship in 1947, Astbury wrote in his reference for Davies: 'He is a very pleasant colleague and is widely cultured in subjects outside his own ... Frankly I do not wish to lose him'.[22]

It was a bad time to be losing such a valued member of staff, for Davies recalled that Astbury had not taken the news of Randall's success with the MRC well:

> I well remember Astbury coming into the laboratory tea that afternoon (it must have been in March 1947) and announcing that 'Randall has been given the MRC grant'. He was a picture of depression.[23]

To have the grand plan for his 'Nunc Dimittis' rejected by the MRC was painful enough, but that they chose instead to support Randall rubbed salt into the wound. In a tribute to Astbury written in 1974, Preston wrote:

> It is all the more tragic to recall that towards the end of his life he became to some degree disillusioned. Apart from the Rockefeller Foundation he had never received much in the way of support either by way of finances or buildings. After the Second World War he saw newcomers in the field supported on a scale he had never received and he felt, it must be admitted with some justification, that the fruits were going to others from the fields he had tilled.[24]

Preston's talk of 'newcomers' was presumably a reference to Randall. But the interest in establishing new centres of biophysics and molecular biology was not confined to the UK. In the years immediately following the end of the Second World War, science fared much better in the USA, which had not suffered bomb damage, rationing, and the general austerity that pervaded the UK. Since the mid-1930s and his early contact with Astbury, Pauling in the USA had become increasingly interested in biology and in particular the structure of proteins. Like Astbury, Randall, and Bernal, Pauling also felt that the time was ripe for an assault on the central questions of biology using the tools of chemistry and physics, and in 1948 he was awarded $100,000 per annum for the next seven years by the Rockefeller Institute to establish a division to do this.[25]

Ever since meeting Astbury back in 1937 and discussing the structure of the keratin chain, Pauling had pondered Astbury's data. During the war years, Pauling had begun to explore the possibility of using a mechanical computing device called a Hollerith machine to perform some of the mathematical procedures that were necessary when interpreting more complex X-ray diffraction patterns. It was a development in which Astbury was particularly interested and he wrote to Pauling asking whether it would be possible to obtain a set of duplicate punched cards for the set of mathematical calculations that Pauling was using.[26-28]

Yet while Pauling and Astbury shared an enthusiasm for applying new techniques such as mechanical computation to the analysis of experimental data, they differed over the interpretation of this data. While Pauling agreed with Astbury that keratin was a long chain of amino acids chemically bonded together, he was convinced that Astbury's proposed alpha form of keratin was wrong in some way, or at least highly oversimplified. Knowing that restrictions on the bond formed between the first carbon atom of an amino acid and the nitrogen atom of its neighbour in the chain meant that these atoms could only lie in a flat plane, Pauling calculated that Astbury's model of alpha-keratin as a chain compressed into a zigzag in two dimensions simply could not work. Applying other chemical considerations such as physical restraints due to hydrogen bonding, he tried to envisage the myriad of possible ways in which the chain might fold up. But one particular detail of Astbury's data always confounded him. All of Astbury's X-ray photographs of keratin showed a spot that suggested some regular structural feature repeating itself along the axis of the chain at a distance of every 5.1 Angstroms. Try as he might, Pauling simply could not envisage how the keratin chain could fold up to give this repeating structure and still obey the laws of chemistry.[29]

It was while on a visit to Oxford as a visiting professor in 1948 that Pauling glimpsed the solution. Lying in bed with sinusitis and bored of reading detective stories, he picked up a piece of paper and began to sketch the angles of the different bonds in the amino acid chain.[30] Then he began to carefully fold the paper into a twisting spiral shape, a helix. The amino acid chain of keratin, he concluded, was folded not in two dimensions, but three—it coiled around into a twisted spiral, or what Pauling called the 'alpha helix'.

Pauling was not alone in trying to solve this problem. He faced competition from Lawrence Bragg's team at Cambridge, who were also using Astbury's photographs of keratin to work out how the chain might fold in three dimensions. But while Pauling was suspicious of the data suggesting that some structure must repeat along the length of the chain every 5.1 Angstroms, Bragg's group used this as the cornerstone for their own model. Both groups postulated that the keratin chain spiralled around its axis in a helix; the difference was in how they interpreted the spot at 5.1 Angstroms on Astbury's photographs.

For Bragg's team, this could only mean that the spiral of the helix went through one complete turn (or 'pitch') every 5.1 Angstroms—hence giving rise to the spot on Astbury's pictures. This would result only if one complete turn of the spiral contained an integral number of amino acids. It never occurred to Bragg's team that Astbury's 5.1 Angstrom spot might well be a distraction arising from some other structural feature, or that the number of amino acids in a complete turn of the helix need not be integral. Not so for Pauling—a combination of the laws of chemical bonding and sheer chemical intuition told him that the 5.1 Angstrom spot was a distraction. He focused instead on showing how a helix in which the spiral repeated every 3.6 amino acids could satisfy all the laws of chemistry.

His hunch was proved right a year later when scientists at Courtaulds, a British firm of artificial fibre manufacturers, reported the successful synthesis of artificial polypeptides. Not only did these artificial peptide chains give X-ray patterns suggesting they were coiled up into a spiral conformation, but also they did not show the 5.1 Angstrom spot that Astbury had obtained when studying keratin chains.[31,32] This single piece of Astbury's data, on which Bragg's team at Cambridge had built all their attempts to understand how the amino acid chain could coil into a helix, arose from a different structural feature entirely. It was Francis Crick who eventually explained why. Crick showed that the polypeptide chains of keratin are coiled in helices, which then coil around themselves forming a kind of 'supercoil'—which repeats every 5.1 Angstroms.[33,34] It was this repeating supercoil that Astbury's X-ray photographs had detected and that had misled the Cambridge team into insisting that the number of amino acids in an alpha helix must be integral.

Pauling and his associate Robert Corey announced their model structure with a short note to the *Journal of the American Chemical Society* in November 1950, with a further seven publications in the *Proceedings of the National Academy of Sciences* during the following year that elaborated on the helical model of protein chains. Pauling's model of the alpha-helical coiling of protein chains earned him the 1954 Nobel Prize in

Chemistry, and his first publication of this work came as an unwelcome shock to Bragg's group at Cambridge, leaving Bragg 'almost apoplectic' that Pauling had got there first.[35]

Nevertheless, the shock of being beaten to the alpha helix did eventually pay dividends for one member of Bragg's team. Max Perutz was so angry and annoyed at having lost out to Pauling that he began to read Pauling and Corey's papers very closely. Perutz noticed that their helical model should give rise to a particular spot on an X-ray diffraction photograph that indicated a regular repeating pattern every 1.5 Angstroms along the length of the keratin chain. At first, Perutz was puzzled why Astbury's photographs of keratin had never shown this spot, but then he realised that Astbury's photographic plates had been too small and that the X-ray beam and fibre had been aligned at the wrong angle to observe this particular spot. Taking a fibre of horsehair, Perutz took a single X-ray photograph and, just as Pauling and Corey's model predicted, there was the 1.5 Angstrom spot. Perutz had obtained the first real evidence that Pauling and Corey's alpha helix occurred in natural proteins. When Perutz first showed the result to Bragg and explained that he had been driven to perform the experiment out of sheer anger at having been beaten by Pauling, Bragg's reply was simple: 'Perutz, I wish I'd made you angry earlier!'[36,37] Perutz's anger was to prove fruitful, for in 1962, with his colleague John Kendrew, he was awarded the Nobel Prize for determining the very first complete molecular structures of the globular oxygen-transporting proteins haemoglobin and myoglobin.[38] What became evident from this work was that the folding of a protein chain into an alpha helix, as predicted by Pauling, was one of the most fundamental and widespread structural motifs in globular and fibrous proteins.

The success of Pauling's work on the alpha helix had exposed the limitations of Astbury's work and left him having to defend his own model for the folding pattern of keratin. On top of this, the rejection by the MRC had come as a blow. He also had to contend with the practical difficulties of establishing and running a new department. The University of Leeds had granted him space to house his new Department of Biomolecular Structure at 9 Beechgrove Terrace. Though now demolished, this was at the time an old Victorian terraced house that stood opposite what is now the Student Union Building at the centre of the campus. Astbury described the premises as 'a makeshift and poor solution' and a far cry from what he had hoped would be a facility in which to carry out cutting-edge science at a bold new frontier.[39] As a former residential property, the premises required a considerable amount of work to be converted into a scientific research establishment.[40] Heating needed to be installed, as did electricity; the floors needed to be strengthened to bear the weight of scientific apparatus; and the house required damp-proofing. In addition, there were issues of drainage, ventilation, and lighting to consider, all of which meant that the premises for the new department were not actually ready for use until 1948.[41]

Yet despite these extensive modifications, there were still problems. There were concerns over the strength of the floors, and the tortuous problems over the installation of a telephone in Astbury's office so that he could speak directly to his X-ray laboratory caused him much frustration.[42] One of Astbury's researchers, Dr Keith Parker, recalled that the electricity supply was unreliable and the floors were uneven, causing the equipment used for delicate spectroscopic measurements to wobble—a problem that was solved by importing a heavy stone bench. But an even greater threat to the delicate equipment was posed by problems with water pressure, which resulted in occasional floods that Astbury feared would damage the X-ray apparatus.[43–45] The premises at 9 Beechgrove Terrace had always been intended as only a temporary home, with the promise of something permanent in the not too distant future, but when it was suggested that Astbury be moved to yet another temporary location while these problems were smoothed out, he expressed his frustration, saying: 'I am simply being squeezed out gradually from one lot of premises anywhere else the University may be able to find for me.'[46]

In addition to these problems, there was a steady succession of minor irritations, all of which distracted from the business of serious research. Having been given a new professorial chair, and elected a Fellow of the Royal Society, Astbury might have expected to be able to devote his time solely to research without the tiresome distractions of the daily logistics that come with running a laboratory. But sadly it was not so, as the following letter to Mrs M. Kelly, Superintendent of the Cleaning Services, shows:

> I am writing to ask if you would now be so good as to include this Department in your regular distribution of clean tea-towels. We have only one tea-towel which is dabbed out from time to time, but I am afraid this

is a very inadequate business, to put it mildly, and we should very much appreciate regular laundering as with the other towels supplied to this Department.[47]

This was just one of the many mundane concerns with which Astbury was distracted. Others included writing to the University Bursar to ask whether the heating could be kept on in the laboratory during the winter, and procuring a decent, reliable typewriter for his secretary.[48,49]

Nevertheless, Astbury refused to let these setbacks and obstacles dampen his enthusiasm for research, and, in the decade that followed the end of the war, he began exploring new avenues of study. One of these was in the new field of electron microscopy, an area that had been brought to Astbury's attention through correspondence with Bernal.[50] The possibility of using focused electron beams to view previously hidden biological structures was a prospect that Astbury said left him with 'bulging eyes and tongue hanging out'[51] (Figure 17). Seven electron microscopes were available to UK facilities as part of the LendLease deal with the US Government. Of these, six were used by the MRC and Manchester, leaving one spare, which Astbury procured for Leeds.[52] Word of this new acquisition caused excitement way beyond academic circles, and Astbury soon found himself inundated with requests from a number of different groups and organisations, ranging from the Institute of Electrical Engineers to the Sixth Form of the Girl's High School at Skipton, all of whom were keen to come and see this new wonder of science.[53,54]

One of Astbury's earliest subjects for study with the electron microscope was the polysaccharide fibres found on the surface of the pneumococcus, the same microorganism that had given Avery his 'bombshell'.[55] But later his main subjects for study were tissue samples from breast tumours and rheumatoid joints.[56–59] This might seem like a sharp departure from studying protein fibres in wool and muscle, but it is likely that his choice of subjects here was highly significant. Both these subjects were of much more immediate medical relevance than his earlier work on wool. His rejection by the MRC had probably been instructive, and he had learned that the way to secure grant funding in future was to emphasise its medical applications and benefits. It was very much a sign of how the landscape of biological research in academia would look in the future.

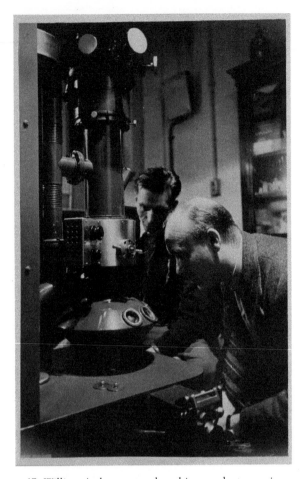

Figure 17 William Astbury at work on his new electron microscope.
Astbury Papers, MS419 Box A.2, University of Leeds Special Collections, Brotherton
Library. Photographer unknown. Reproduced with kind permission of the University
of Leeds, Brotherton Library Special Collections.

It was a period of contrasting fortunes for Astbury. He had suffered
rejection by the MRC and seen his model for keratin come under attack by his old friend Pauling, as well as struggling with the challenges
of establishing a new department. But it would not be strictly accurate
to portray him as struggling along 'cap in hand' while Randall, Pauling,
and Bernal were awash with funding. Although his financial situation
may have been initially tough, by 1955, he had secured substantial

funding from bodies, including the Rockefeller Foundation, Imperial Chemical Industries, and the Nuffield Foundation.[60] He had also acquired numerous honours and accolades during this period, including an invitation to attend a garden party at Buckingham Palace, an honorary degree from the University of Strasbourg, and an award from the Swedish Royal Society of Science. In 1950, he was made a member of the New York Academy of Sciences and three years later received a medal from the American Society of European Chemists.[61]

But despite all these accolades, the one goal that he had really hoped to achieve had eluded him. He had failed in his plan to make Leeds the national centre of molecular biology. The attention of the MRC was now on London and Cambridge, not Leeds. Perhaps this is why in *Dark Lady of DNA*, her biography of Rosalind Franklin, the writer Brenda Maddox described Astbury as an 'unpersuasive man at an unfashionable northern university'.[62] It is a charge that is perhaps somewhat unfair. For while Astbury may well have failed to persuade the MRC to support his own vision for a centre of molecular biology, his lasting influence was far from insignificant, and science in Leeds was yet to play a vital role in the story of DNA.

When John Randall had originally submitted his proposal to the MRC in 1946 to set up a new unit at King's College to study what he called 'biophysics', he credited Astbury as one of the main people with whom he had discussed the project and its proposed research aims.[63] In his proposal, Randall declared that the aim of the unit would be the direct and indirect study of biological material by physical methods. Examples of the projects that he proposed included the study of chromosome breakage and mitosis using ultrasound, the use of nematode eggs to study the formation of the spindle structure that pulls dividing chromosomes apart during cell division, the study of cytoplasmic streaming, and the use of tracer elements to follow biochemical processes. The only reference to X-rays was a proposal to use them to study the structure of the heads of sperm cells, and there was certainly no direct reference to using X-ray diffraction to study the structure of DNA.[64]

Yet it was in Randall's Biophysics Unit at King's that only a few years later Franklin would use X-ray diffraction studies of DNA to obtain the famous 'Photo 51'.

The impetus for the group at King's to begin studying the structure of DNA had actually come not from Randall, but from his right-hand man, the biophysicist Maurice Wilkins, who had worked with Randall at both Birmingham on the cavity magnetron and later at St Andrews, before accepting a post in Randall's new unit in London. During the Second World War, Wilkins had worked on the Manhattan Project to develop the first atomic bomb. But following the horrific devastation of Hiroshima and Nagasaki, he began to feel uneasy about nuclear physics, particularly the enthusiasm that some scientists showed for developing the more powerful hydrogen bomb. When the first bomb was dropped on Hiroshima, Wilkins was working in San Francisco and recalled scenes of jubilation at the prospect that this must surely herald the end of the war. However, when a friend that evening remarked, 'This is Black Monday—I always hoped it would never work', Wilkins said that he felt rather small and could only respond 'Yes, you are right.'[65]

That short reply marked the end of his career as a nuclear physicist, but a change of direction came when he read Schrödinger's *What is Life?*[66] What excited Wilkins about the book was that it showed that a physicist could bring important insights to bear on biological systems, and it left him convinced that here was an area for future research.[67] Following his reading of Schrödinger, Wilkins began to take an interest in genetics, and, on learning of how the geneticist Müller had successfully induced mutations using X-rays, he attempted to do the same with ultrasound. When this proved to be unsuccessful, he turned his attention to developing optical microscopes that used visible, UV, and infrared light to study the movement of chromosomes and DNA within the cell, by which time he had been appointed Assistant Director of Randall's Biophysics Unit.[68]

It was also during this period that Wilkins learned of Avery's work showing the importance of nucleic acids in pneumococcus. Like Astbury, it left him feeling excited and convinced that he was on the right track in choosing to study DNA. But it was Gerald Oster, a visiting American scientist, who suggested to Wilkins that perhaps using microscopes was not the best approach.[69] Oster had been working on using X-ray diffraction to explore the structure of the tobacco mosaic virus, an area of work first begun by Bernal, Astbury's old colleague from his days with William Bragg, and he suggested that Wilkins use the same approach to study the structure of biological molecules, rather than just follow their movements around the cell.[70]

In his autobiography, *The Third Man of the Double Helix*, Wilkins acknowledged that one figure had already blazed the trail for him in this area.[71] It was Astbury's X-ray work at Leeds, he said, that had begun to address the mystery of the structure of DNA, and, thanks to these pioneering studies, Wilkins was convinced that X-ray diffraction was the route that he should follow.[72] At a conference in London, he met the biochemist Signer from Berne, who had developed a method of extracting DNA that left it in pristine condition very similar to that in living cells. Signer made a generous offer to give this material to anyone who was interested, and, when Wilkins obtained a sample, he and a PhD student called Raymond Gosling began to study it using X-ray diffraction.[73]

The diffraction patterns obtained by Wilkins and Gosling were beautiful. Although Astbury and Bell's X-ray work had confirmed that DNA was a chain and provided important information about the spacing between bases, they could give little more information. What Wilkins and Gosling had realised was that the DNA was highly sensitive to its environment and in particular the humidity. By preventing the samples from drying out too much, they obtained crystalline diffraction patterns showing far more spots than Astbury's early pictures and so potentially giving much more information about the molecule.[74] When Wilkins presented the pictures at a conference at the Zoological Station in Naples in the Spring of 1951, Astbury, who was present in the audience, congratulated him on a set of beautiful results.[75] It was the first time that Wilkins had met Astbury, and he recalled that:

> he seemed genuinely warm when he said our pattern was much better than anything he had got. I was very glad to meet him because he seemed a real human being . . . My friend the Cambridge chemist John Kendrew, told me he got to know Astbury best over a bottle of whiskey when Astbury discussed his love of playing the violin.[76]

Encouraged by Astbury's support, Wilkins returned from Naples eager to continue his work.[77] But there was a problem. The relationship between Wilkins and Randall was not always a smooth one. Wilkins once described Randall as having a 'Napoleonic' style of leadership, and he could sometimes make life difficult for Wilkins to the point that the two of them occasionally had stand-up rows.[78,79] Speaking of Randall in his autobiography, Wilkins said: 'I admired and respected him, but I cannot really say that I found him very likeable.'[80] And there was probably no greater example of the difficulties between the two men than Wilkins' work on DNA.

It had become apparent to both Wilkins and Randall that the Biophysics Unit at King's needed a real expert in X-ray crystallography, and in 1950 they found one. Franklin was currently finishing some research in Paris and accepted the post at King's with the intention of starting in January 1951. Randall had initially intended for her to use X-ray diffraction to study the denaturation of proteins in solution, but, towards the end of 1950, he wrote to her with a very different plan in mind, suggesting that 'it would be a good deal more important for you to investigate the structure of certain biological fibres in which we are interested'.[81] Furthermore, in the same letter, Randall went on to say that 'as far as the experimental X-ray effort is concerned there will be at the moment only yourself and Gosling . . .'.

It was Wilkins who had actually suggested to Randall that Franklin be given the task of using X-ray diffraction to study the structure of DNA and that his PhD student Raymond Gosling be transferred to her.[82] It made sense—she was, after all, the expert in X-ray crystallography, and so was best placed to teach Gosling the research student his trade. But what Wilkins never anticipated was that Randall's poorly worded letter would give Franklin the mistaken impression that X-ray research into the DNA structure now belonged solely to her, a mistake that was reinforced when Randall held a meeting with Franklin in his office shortly after her arrival in January 1951 while Wilkins was away on holiday.[83]

The arrival of Franklin at King's should have resulted in a fantastic collaboration with Wilkins and might even have resulted in them solving the structure of DNA. But, thanks to Randall's poor leadership and lack of clear communication, it was destined not to be, and their relationship deteriorated to the point where they were barely speaking. It was against this background of strife that Franklin obtained 'Photo 51' the following year, but the origins of the X-ray work at King's can all be traced to the influence of Astbury, and it was to him that Randall, now very much interested in the structure of DNA, turned when wishing to discuss the matter.[84–86]

Astbury's 'Nunc Dimittis' may have gone unsung, but, thanks to the influence of his work on Wilkins and the group at King's, he had ensured that his research had played a key role in the discovery of the double helix. And nor did the role of Leeds in the story of DNA end with the failure of Astbury's own plans for a national centre for molecular biology. For while it may well have been out of fashion with the MRC, Astbury's adopted home city and its wool industry was yet to play another invaluable role in the discovery of the double helix. Research into textile fibres had been the cornerstone of Astbury's success that had led him to found molecular biology. And it also gave birth to an innovation that would revolutionise this new science.

After spending his mornings at home dealing with paperwork and administrative tasks, Astbury would stroll from his home in Headingley down to his laboratory on the University of Leeds campus in the early afternoon. Halfway down Headingley Lane, he would pass 'Torridon', a grand old Victorian house that had been converted into the technical headquarters of the Wool Industries Research Association (WIRA) and where, working in a former stable now billowing with chloroform fumes, the scientists Archer Martin and Richard Synge were developing a new method for the chemical analysis of the keratin proteins in wool. Known as partition chromatography, Martin and Synge's method was based on a very simple principle that can easily be demonstrated with a strip of filter paper, some felt-tip pens, and a glass of water. A series of dots are made with different-coloured pens along the width of the filter paper, the end of which is then dipped into the glass of water. As the water is drawn up into the paper, the various pigments in each spot are carried up the paper by the water, with each colour moving a different distance depending on the solubility of its ink in the water. Martin and Synge applied this same principle to develop apparatus that enabled the separation and quantification of the individual amino acids that made up the polypeptide chains of the keratin proteins in wool.

The impetus for Martin and Synge's work had come from the International Wool Secretariat (IWS), a body which had been funded by the wool growers of Australia, New Zealand, and South Africa for research into wool.[87] Hedley Marston, an Australian advisor to the IWS, had studied the effect of sulphur-deficient soils on sheep's wool. He worked in same laboratory in Cambridge where Synge was investigating the separation of amino acids according to their respective solubilities. One of the concerns of the IWS was that wool might well be displaced by newly emerging artificial fibres, and Marston urged that this be addressed by directing fundamental research into the chemical nature of wool.[88] In particular, he recognised that Synge's work might be of great benefit to such work and suggested that he be provided with a studentship to analyse the amino acid composition of wool.

To help Synge tackle this problem, he was introduced to Archer Martin who was an expert in the field of chemical separation. The question of how to separate and purify compounds from mixtures was one that had preoccupied Martin for a long time. While still at school, he liked to solder together empty old coffee tins and fill them with coke to make his own five-foot high reflux distillation columns.[89,90] Whilst working at the University of Cambridge he had put these skills to good use building a purification apparatus, before bringing it with him to the WIRA labs in Leeds in 1938.

Three years later, Synge also moved to Leeds to join WIRA and the combination of his knowledge of protein chemistry with Martin's practical expertise resulted in a method that would leave a lasting legacy that went far beyond the analysis of wool.

Astbury was no stranger to Martin and Synge's work and collaborated with them. In 1941, before she left Leeds for war service, Florence Bell had carried out a number of X-ray studies of small molecules for Martin and Synge and appeared with them on a paper published that year. And although he disagreed with Martin and Synge on some important aspects of protein structure, Astbury recognised that their development of partition chromatography had tremendous potential, for, in a letter to Kenneth Bailey, he wrote: 'I had a look the other day at Martin and Synge's partition chromatography machine and it thrills me a lot.'[91,92] Equally thrilled was the Cambridge biochemist Fred Sanger. Astbury had shown that keratin was a long chain of amino acids, but Martin and Synge wanted to find a way of separating these amino acids and determining the precise order in which they were arranged. The problem was that the keratin proteins in wool were simply too big. So Sanger applied Martin and Synge's method to a much smaller protein—the hormone insulin—and was able, for the very first time, to determine the complete order of amino acids in a protein, for which he was awarded the 1954 Nobel Prize in Chemistry.[93,94]

Sanger's work revealed that the amino acids which make up a protein chain are arranged in a precise linear order, or sequence, and it is this that makes one protein such as insulin, different from another, for example, haemoglobin. This landmark insight transformed our understanding of life at the molecular level and Sanger always acknowledged

Figure 18 The two sole remaining features of the Wool Industries Research Association (WIRA) at 'Torridon', a house that once stood on Headingley Lane, Leeds, UK where Martin and Synge developed partition chromatography. (a) The ruin of the building, a former stable, that once housed Martin and Synge's laboratory (b) A gatepost bearing the name 'Torridon (c) Plaque on the gatepost unveiled by the Leeds Philosophical and Literary Society commemorating the award of the 1952 Nobel Prize in Chemistry to Martin and Synge's work for the work that they did here.

Photographs by K.T. Hall and Leeds Philosophical and Literary Society.

that it could not have been done without Martin and Synge's method of partition chromatography, for which they themselves had been awarded the 1952 Nobel Prize in Chemistry.

To do full justice to the impact of Martin and Synge's work probably deserves an entire book itself,[95] but as part of its bicentenary celebrations in 2019 the Leeds Philosophical and Literary Society unveiled a plaque in their honour on the gatepost that still bears the name 'Torridon' and which stands on Headingley Lane (Figure 18). This, along with the solitary ruin that was once Martin and Synge's lab, is now all that remains of the grand old Victorian property that once housed the WIRA facility and today this part of Leeds is better known for being the route taken by hordes of weekend revellers in fancy dress as they partake in a long established student pub-crawl, rather than Nobel-Prize winning science that changed the world.

For change the world it did. The thrill that Astbury described in his 1941 letter to Bailey at seeing Martin and Synge's chromatographic apparatus came from his realisation that this opened up a vista of new possibilities for research into proteins—possibilities that Sanger had quickly realised. However, on the other side of the world in New York, Erwin Chargaff, another biochemist, had already realised that Martin and Synge's work could be applied far beyond proteins and might well help in solving the secrets of DNA.

9

'One Grand Leap . . . Too Far'

Despite being one of its founding fathers, US biochemist Erwin Chargaff actually had a very dim view of molecular biology, saying of its practitioners:

> That in our day such pygmies throw such giant shadows only shows how late in the day it has become.[1]

Rather like Baron Frankenstein in Mary Shelley's famous cautionary tale, Chargaff came to fear that he had spawned a monster, which was evident from this rather sour and caustic pronouncement on the entire discipline.

He was born in 1905 in Czernowitz, a provincial capital of the Austro-Hungarian Empire that later became part of Ukraine. In 1914, when Czernowitz was occupied by the Russian Army, his family moved to Vienna, where he studied chemistry at the University there. In 1928, he took up a research post at Yale in the USA to work on the chemical composition of the tubercle bacterium. But he was uncomfortable in the USA, complaining about settling in a country that was 'younger than most of Vienna's toilets' and, in 1930, moved to Berlin.[2] With the rise of fascism across the continent, this was hardly the best time to be returning to Europe, and in his memoir *Heraclitean Fire*, Chargaff described his move as 'a rare case of a rat returning to a sinking ship'.[3] Three years later, he moved to Paris to work on the Bacillus Calmette–Guérin (BCG) vaccine, but as the situation in Europe became increasingly ominous, he left once again for the USA, where he finally settled in the Department of Biochemistry at the College of Physicians and Surgeons at the University of Columbia in New York City.

Until now, most of Chargaff's work had been concerned with the biochemistry of bacterial lipids, but he had also done some work on nucleic acids while working for the US Army Medical Service on Rickettsia during the Second World War. News of Avery's paper on the role of nucleic acids in pneumococcus, together with Schrödinger's *What is Life?*, had a profound effect on him. Schrödinger's book proposed that a sufficiently large molecule with enough variation in its structure might act as a 'code-script' to carry hereditary traits and thanks to Avery, Chargaff began to wonder whether nucleic acids might fit the bill better than proteins:

> I saw before me in dark contours the beginning of a grammar of biology ... Avery gave us the first text of a new language, or rather he showed us where to look for it. I resolved to search for this text.[4]

It was the work of John Mason Gulland, an English chemist that pointed the way for Chargaff to begin his search. Gulland held the Sir Jesse Boot Chair of Chemistry at the University of Nottingham where his main interest was in the chemistry of nucleic acids. In 1945, he suggested revising the model of DNA made of repeating groups of four nucleotides, or 'tetranucleotides'. Rather than being made of physical tetranucleotide units, Gulland proposed that a DNA chain might be composed of what he called a 'statistical tetranucleotide'; in other words, the molecule contained equal amounts of all four bases, without them necessarily occurring in the same sequence of four over and over again. In this case, two DNA molecules might contain identical numbers of the four bases but differ according to the *sequence* in which those bases occurred.[5]

It was a profound insight, and one that was highly influential on Chargaff. He wondered whether this difference in the sequence of bases might account for the ability of nucleic acids to carry species-specific properties, as Avery had shown in his work with pneumococcus.

Together with the publication of Schrödinger's *What is Life?*, this came as a pivotal moment for Chargaff, after which he devoted all his research to the study of nucleic acids. After attending a meeting at Cold Spring Harbor in 1947 at which Gulland presented data on the proportions of the four bases in DNA, Chargaff decided to focus his efforts on making accurate measurements of the bases adenine, guanine, cytosine, and thymine in nucleic acids from various sources.[6] According to Levene's tetranucleotide hypothesis, all nucleic acids should contain exactly the same amount of each base, but this supposition was derived from purely theoretical calculations. Chargaff wanted to test this idea empirically by separating out the four individual bases of which DNA was composed and quantitatively measuring their respective amounts.

But this presented a formidable practical obstacle, as thus far, there was no reliable method of separating and measuring small compounds such as the bases in DNA in tiny quantities.

Chargaff's grand plan to unravel the secrets of DNA appeared to be at a dead end until Archer Martin visited Columbia to give a partition chromatography lecture, in which he described a refinement of the method using filter paper.[7] Martin's lecture was a gift to Chargaff.[8] It was just what he needed for his proposed plan to analyse and quantitate the base composition of nucleic acids. Chargaff realised that, although the scientists at Torridon had applied their chromatographic techniques to analysing the amino acid composition of keratin protein chains in wool, the same method could equally well be adapted for analysing and measuring the bases in a nucleic acid chain, and in a series of papers published in 1948 he proved that this was possible.[9–11] There now remained the question of whether the biological specificity of the nucleic acids occurred at the level of their base composition. If nucleic acids were indeed responsible for determining the various biological traits that distinguished one species from another, as Avery suggested was the case in pneumococcus at least, and if this species specificity resided at the level of differences in base sequence, then two predictions could be made. Firstly, in nucleic acids isolated from different tissues of the same species, these four bases should occur in identical proportions, and secondly, their proportions should vary between nucleic acids extracted from different species.

In order to test the first prediction, Chargaff analysed the base composition of DNA extracted from calf thymus and spleen; as expected, he found that there was no significant difference in the base composition.[12] The question now remained of whether nucleic acids from different species showed a variation in their base composition. To address this issue, Chargaff analysed nucleic acids from two different species of microorganism—avian tubercle bacillus and yeast.[13] The results showed that, sure enough, there was a significant difference in the base composition of nucleic acids extracted from these two microorganisms, suggesting that the differences between yeast and the avian tubercle bacillus somehow originated within the sequence of their respective DNA molecules.

In the following year, Chargaff extended this work to include human DNA and he also speculated on how exactly variations in the base composition of nucleic acids might result in distinct biological traits that distinguished one species from another.[14] Chargaff's suggestion was that differences in base composition might have 'far reaching changes in the geometry of the conjugated nucleoproteins'.[15] In other words, the varying sequence of bases might produce changes in the physical shape of the DNA molecule.

Chargaff's suggestion that different sequences of bases might result in structural variation that could represent biological traits fired Astbury's imagination. Only a few years earlier, in 1947, he had presented new X-ray photographs of DNA, the significance of which is open to interpretation. Some have described Astbury's view of the function of DNA at this point as representing a 'giant step backwards' towards the inert model of a chain of repeating tetranucleotides.[16] Yet even if the majority of the nucleic acid chain was a repeating tetranucleotide with some 'more standard function', Astbury had suggested that there might still be some regions in the chain where the base sequence varied in a way that generated changes in its shape.[17] Now Chargaff seemed to be proposing exactly the same thing. If differences in the base composition of nucleic acids from different species did indeed cause large-scale structural changes in the shape of the DNA chain, then X-ray diffraction was the ideal method to detect them. By studying structural changes in the DNA molecule as revealed by an X-ray diffraction photograph, it might well be possible to explain how DNA could carry biological information of the kind that Avery had observed in his experiments on pneumococcus.

The secret of DNA seemed to be within Astbury's grasp, and in 1950 he wrote to Chargaff asking whether he could send some samples of nucleic acid for analysis by X-ray methods:

> What about the specimens of very pure and highly polymerised nucleic acids which you kindly offered to send me when we had our pleasant little talk in your lab last summer. I should very much welcome the opportunity of seeing what can be found out from them by X-rays, especially since I have been re-reading your communications to Nature of May 13 1950. The differences in purine and pyrimidine composition that you report should be clearly demonstrable in X-ray diagrams if only we can have preparations that can be highly oriented enough.[18]

In the Spring of 1951, Chargaff sent his promised samples to Astbury, who passed the samples on to his research assistant Elwyn Beighton and gave him the task of analysing them by X-ray diffraction[19,20] (Figure 19).

Figure 19 Astbury's research team in his Department of Biomolecular Structure. Photographer unknown. Top row, left to right: Unknown, Unknown, Spark, Ronald Reed, Alfred Millard, Ian McArthur, Elwyn Beighton. Bottom row, left to right: Kenneth McLaurin Rudall, William T. Astbury, H. J. Woods. Photograph property of Dr K. D. Parker. Image of Beighton reproduced with kind permission of Mrs S. Sanderson and Ms G. Sanderson.

Having joined Astbury's laboratory in 1939 as 'lab-boy', this was certainly a far cry from Beighton's original duties, which would have ranged from shifting equipment around the laboratory to making the tea. Keen to improve his lot, Beighton had then become a technician while at the same time studying for a BSc under a scheme that was then offered by the University of Leeds and gave technical staff the opportunity to study for a degree.[21]

During the Second World War, Beighton was called up for service, and, just as he had done for Florence Bell, Astbury wrote a letter requesting that Beighton be allowed to remain in his laboratory on the grounds that he was a valued member of staff and that the laboratory was doing work on fibres that was of use to the government.[22] Again, just as had happened with Bell, Astbury's letter was in vain and Beighton spent the war applying his scientific training to radio work. During the war, Beighton kept in touch with Astbury, writing to him to ask for his support in applying for a commission in the RAF.[23] As the war ended, Beighton wrote to Astbury asking for a different kind of support. He had heard of Astbury's plans for a new department of molecular biology, and he wondered whether there might be a job for him in this new facility. In his letter to Astbury, Beighton said that, while his radio work during the war had been enjoyable, he found that 'the juggling of resistors and condensers when the apparatus involved is an end in itself has always seemed a very cold occupation compared with the design and use of such apparatus in research especially in molecular structure'.[24] Beighton's ambitions had now gone beyond being a technician. He described feeling a 'thrill' at the thought of research, and wanted to be part of Astbury's grand new project, saying: 'It is a great honour to be considered worthy by you to assist in the building up of your new Department'.[25]

When he had sent his samples to Astbury, Chargaff had described them as a 'trial balloon' and apologised that they might not give the best X-ray diffraction photographs on account of their low water content.[26,27] His concern and apologies were unnecessary. To prepare the material for analysis, Beighton stretched the fibres by 300% and dripped water onto them to keep them sufficiently hydrated.[28] Then, using an improved X-ray machine in which the anode rotated so as to minimise damage caused from the impact of the electron beam, he took a series of 15–20-minute exposures of the fibres. The resulting photographs were a vast improvement on those taken by Florence Bell back in 1938, for, having taken care to ensure that his material was sufficiently hydrated, Beighton was able to work with samples of the pure 'B' form. His efforts were certainly rewarded. When the resulting photographic plates were developed, they all contained one striking feature: an arrangement of black spots at the centre in the shape of a letter 'X' (Figure 20). In a letter written in 1992, historian Robert Olby described Beighton's photographs as being 'clearly the famous B-pattern found by Rosalind Franklin and R. Gosling'.[29] Yet while Franklin's 'Photo 51' made James Watson's jaw drop and set his pulse racing, Astbury's response to Beighton's photographs could hardly have been more different. Whereas Franklin's photographs drove Watson to sit scribbling the details on his newspaper as he rode on the train home from London to Cambridge, Astbury never published Beighton's work in a scientific journal, or even presented it at a meeting. Rather than become the crowning achievement of Astbury's scientific career, this was the last piece of work that Astbury ever did on DNA. Having pioneered the use of X-rays to study nucleic acids and published the first X-ray diffraction photographs of their structure, Astbury now fell suddenly silent on the subject of DNA and never returned to it.

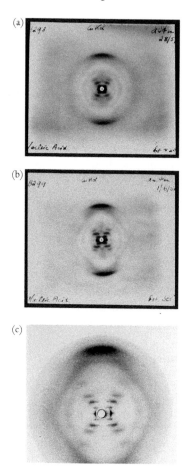

Figure 20 X-ray diffraction photographs taken by Elwyn Beighton in Astbury's laboratory of 'B'-form sodium thymonucleate on 28 May 1951 (a) and 1 June 1951 (b). A striking feature of both patterns is the same central black cross pattern seen in 'Photo 51' taken by Rosalind Franklin a year later (c). (a, b) Astbury Papers, MS419 Box C.7, University of Leeds Special Collections, Brotherton Library. Reproduced with the kind permission of the University of Leeds, Brotherton Library Special Collections. (c) Reproduced with the kind permission of King's College Archive.

Writing nearly 40 years after Beighton's photographs were taken, Astbury's colleague Mansell Davies said that 'for so bright a light in the

DNA field to have gone out so suddenly' was a mystery that demanded an explanation.[30] And Davies believed that he could offer at least a partial explanation. The rejection of Astbury's proposal for funding of his new department by the MRC had taken its toll and, according to Davies, this was a blow from which he never truly recovered:

> Is it too much to suggest that after his entirely reasonable expectations took a very heavy blow from the MRC, Astbury's enthusiasm deserted him and he lost heart for further demanding effort? Certainly his periods of ebullience were accompanied by spasms of depression.[31]

According to Davies, in the years that followed his rejection by the MRC, Astbury became disheartened, and the passion and fire that had characterised his early work was quenched. Davies suggested that perhaps the pivotal moment for Astbury came early in 1951 when he attended the conference at the Zoological Station in Naples at which Maurice Wilkins presented his X-ray diffraction photographs of 'A'-form DNA. While Wilkins' pictures inspired Watson to solve the structure of DNA, Davies suggested that they had a very different effect on Astbury, who now conceded that research into DNA belonged to better-funded groups such as Randall's Biophysics Unit.

Davies was not alone in observing that, towards the end of his career, Astbury seemed to lose heart. Reginald Preston, Astbury's long-time friend and colleague with whom he had studied the cell wall of *Valonia*, said 'It is all the more tragic to recall that towards the end of his life he became to some degree disillusioned.'[32] According to Preston, the cause of this disillusionment was that Astbury now saw newcomers to the field of molecular biology being supported on a scale that he had never known.[33] After all his hard work, Astbury felt that he had still never received the support that he felt he deserved from UK bodies and had been forced instead to rely on funding from foreign institutions such as the Rockefeller Foundation. It was a feeling that was brought into sharp focus after his rejection by the MRC and one that was evident in a letter written in 1945 to Sir Charles Martin at Cambridge:

> these last few years I have often felt pretty depressed that what I had considered so wonderful an intellectual adventure, and so worthy of generous support, should make so little impression here.[34]

There were also deeper, more existential reasons for the lack of the exuberance that had characterised Astbury's earlier years. In his later life, colleagues noted that 'he seemed to undergo a phase of philosophical questioning and perplexity', and it was perhaps only half in jest that he said he would rather chat to his young grandson while walking together through the streets of Headingley than engage in conversation with his professional associates.[35,36] It was a period in which he seemed to question the worth of all his efforts over the previous decades. In a letter to Jerome Alexander, an industrial chemist, Astbury declared:

> I should warn you though that as the years roll by, I yet am afraid that I get less and less satisfaction out of science myself . . . I get a bit depressed sometimes at the seeming utter insignificance of almost anything we do against the backdrop of the endless cycles of creation, and the way the 'solution' of one problem only serves to lead to the next and I get that awful craving to be vouchsafed a glimpse, if only the tiniest, of what it is all about. 'And all our yesterdays have lighted fools the way to dusty death!'[37]

Amid this general feeling of weariness, there was one source of solace: writing to Pauling, Astbury described music as being his 'chief consolation in a very tiresome world'[38] Music had always been a source of inspiration to him, but in this last part of his life, 'he came in part to feel that much of the time he had spent on science might with more spiritual profit have been devoted to music'.[39] One of the greatest moments of his life had been to have dinner with the famous violinist Yehudi Menuhin, and it was now that he first began learning to play the instrument himself—perhaps as a refuge from the disappointments in his professional life.[40] Preston recalled that Astbury became 'a passable player even though his finger tips on the left hand were continually bleeding or calloused' and that the two of them spent many joyous evenings playing duets—Preston on the piano, Astbury on the violin—right up until only a few months before Astbury's death.[41]

It was also to music that Astbury had once turned when searching for a metaphor to describe the main theme of his scientific work: that the fibrous chain molecule was 'Nature's chosen instrument in the symphony of Creation'.[42]

His twin passions for classical music and molecular structure found poetic expression in an X-ray photo that Astbury presented in 1960 to the Jubilee Conference of the Textile Industries.[43] The image (Figure 21) had been taken by Beighton two years earlier and showed the X-ray

(a)   (b)

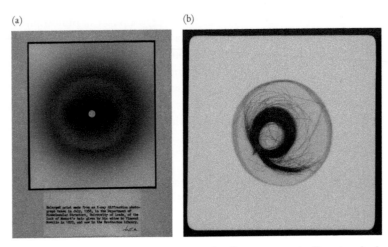

Figure 21 a) X-ray diffraction photograph of keratin protein fibres in a lock of Mozart's hair (b) that was found within a set of travel diaries kept by the composer Vincent Novello and which were acquired by the University of Leeds in 1950. The X-ray image was taken by Astbury's research assistant Elwyn Beighton in July 1958 and presented by Astbury at the Jubilee Conference of the Textile Industries in 1960 (Journal of the Textile Industries, 1960; p.525). University Archive Collection LUA/PHC/002/12 Reproduced with the permission of Special Collections, University of Leeds Library.

diffraction pattern of keratin fibres in a lock of hair taken from Mozart's own head.[44] With his own death only a year away, perhaps the tears that Astbury is said to have wept at the sight of this image were born of his growing sense of disillusionment, of time running out, and glorious opportunities missed. They may equally however have been tears of joy. For despite the ever-growing weight of disillusionment and loss of heart described by Davies and others, one passion that remained undiminished was his conviction that within the shape of these chain molecules lay the secret to understanding living systems.

Having used X-rays to explore the structure of wool, muscles, and feathers, he had by now become interested in a new kind of biological fibre that excited him so much that he said he could hardly sleep at night.[45] The source of this excitement was the whip-like appendages found in bacteria called flagella, by which individual organisms can propel themselves through water (Figure 22). The samples of flagella had been obtained from the common bacteria *Proteus vulgaris* and *Bacillus subtilis* by gently shaking a bacterial suspension until the main body of

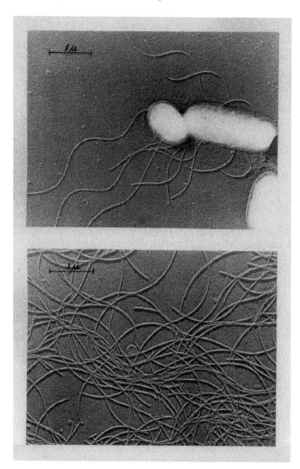

Figure 22 Electron microscope images of bacterial flagella taken by Elwyn Beighton. In 'X-Ray Studies of the Structure of Bacterial Flagella'. PhD thesis, University of Leeds, 1952. University of Leeds Library; Shelfmark THESES S/BEI, pp. 29–30. Reproduced with kind permission from the University of Leeds Library.

the cell fell off. The separated flagella were then separated from the bacterial cells by sedimentation in a centrifuge. This method had been developed by a team of researchers in Uppsala, Sweden and when one of the group, Dr Claes Weibull, visited Astbury's laboratory in 1948, he brought a sample of the material for Astbury to investigate by X-ray diffraction.

In a lecture called 'How to Swim with a Molecule for a Tail', delivered at Wayne State University, Michigan in 1953, Astbury's excitement was palpable. Lighting up a cigarette to help him 'talk more intelligently', he presented his audience with X-ray diffraction patterns obtained from the flagella of bacteria isolated from the teeth of one of his own laboratory members.[46] What excited him to the point of sleepless nights about the flagella was that the X-ray analysis showed that their protein chains underwent the same kind of transformation from a compacted to an elongated form that he had observed for keratin proteins in wool and myosin proteins in muscle. Writing to Professor Archibald V. Hill, an eminent pioneer in muscle physiology at University College London, Astbury described this result as 'thrilling' because it showed that bacterial flagella, muscle myosin, and wool keratin all belonged to the same molecular family.[47] In other words, while living systems such as a muscle and a bacterium might at first sight appear to have little in common, this appearance was deceptive. For, beneath this apparent superficial diversity, there was an underlying unity. Speaking on BBC radio in 1957, Astbury explained that throughout living systems, there lay the 'unifying idea of stretchability and contractility, elasticity and motility, based on lengthening and shortening, the unfolding and folding of polypeptide chains'.[48] For Astbury, living systems were characterised by movement, and this was true right down to the molecular level. Bacterial flagella, he argued, could be thought of as a kind of 'monomolecular muscle'.[49]

Astbury described his work on flagella as having 'afforded us more molecular biological joy, I should say, than almost anything else I can remember since those first great days with the alpha–beta transformation'.[50–52] This idea of a unifying theme within nature, based around changes in molecular configurations, had defined his career and brought him success and renown on the international scientific stage. But it also proved to be the stumbling block that cost him the prize of the double helix and might well provide a more complete answer to Davies' question as to why, despite having such a vital clue in his grasp, Astbury did nothing with Beighton's photographs.

Despite feeling that Astbury's flagging spirits had taken their toll on his scientific work, Davies still refused to accept that an X-ray crystallographer of Astbury's stature would have overlooked the fact that the patterns in Beighton's photographs had some clear significance. It has been suggested that, as a physicist, Astbury was never really interested in biology, but rather was concerned with applying methods and techniques that could reformulate biology in terms of physics simply for the sake of doing so—almost an act of intellectual imperialism on the part of physics to subjugate biology.[53] Yet Astbury's heart was much more inclined towards biological problems than pure physics, for, as he once remarked to Randall, 'I know a fair amount about molecules, and especially biological molecules, but am not at all clear about what happens inside a wireless set.'[54] In addition, the excitement of his response to Avery's experiments reveals that biological questions, particularly the function of nucleic acids, were of prime importance to Astbury, and therefore for him to have paid no attention whatsoever to Beighton's photographs was, in Davies' words, something akin to 'the suggestion that the manager of a football team does not appreciate it when his side scores a goal'.[55],[56]

Davies' question was also echoed by Beighton himself. Having obtained a PhD under Astbury's supervision, Beighton went on to become an academic at Leeds in the area of biophysics, and, though he rarely talked about his work with his family, there were occasional moments when he betrayed to them a sense of frustration that Astbury had never grasped the full implications of his X-ray photographs.[57] Yet Beighton's own PhD thesis provides a clue as to why he never did so. The subject of Beighton's doctoral research was X-ray studies of bacterial flagella. In the course of his work, Beighton outlined how the whip-like motion of flagella might be the result of an alternating wave of supercontraction and elongation in the molecular chain that propagated along its entire length.[58] It was that the motion of flagella could be explained in terms of a change in molecular configuration that so excited Astbury. Flagella were, he said, a 'prototype of the muscle machine; the mechanism stripped to its barest essentials; a capacity for rhythmic, energetic movement embodied in a single molecule'.[59] The secret of life was to be found in movement at the molecular level. And that, for Astbury, was the problem in understanding DNA and how it worked.

Davies said he was baffled as to why Astbury never published Beighton's pictures, 'even if, surprisingly he did not grasp the helical structure revealed by the striking cross-pattern'.[60] But perhaps Davies was asking the wrong question. His surprise that Astbury failed to spot a helix from the patterns of spots in Beighton's pictures is surely a classic case of being wise after the event. Rather than ask why Astbury never spotted a helix, it might be more revealing to ask what *did* he see in the pattern of spots in Beighton's photographs? And when Davies pointed out that even a novice in X-ray diffraction looking at Beighton's pictures ought to have been able to spot that 'the sharpness of the spots and the simplicity of the patterns now instantly suggest that some well-defined and highly regular structure was present in the specimen', he had himself unwittingly provided an answer.[61]

When Watson saw Franklin's photographs, he immediately recognised that the cross pattern spoke of a chain molecule with a 'well-defined and highly regular structure'. Astbury may well have also come to the same conclusion. The crucial difference lay in how Watson and Astbury interpreted the meaning of this regularity. For Astbury, the structure was *too* well-defined, *too* highly regular—there was no sign of the variation in shape or molecular movement that he believed was the basis of living systems and that characterised keratin in wool, myosin in muscle, and the protein fibres of bacterial flagella. In actual fact, DNA *did* exhibit a conformational change. Wilkins had showed that DNA fibres could be stretched in a way that was reminiscent of the alpha-to-beta transition in keratin, but, whereas this transition in keratin explained the elasticity of wool, in the case of DNA, stretching the fibres simply resulted in what Wilkins described as an 'irregular muddle'.[62] Franklin had shown by discovering the 'A' and 'B' forms of the molecule that its conformation could change depending on water content. But this was a much more limited change in conformation and one about which Astbury was unclear. Even if his X-ray studies had been able to detect this particular transition, it was nowhere near as striking as the alpha-to-beta transition of keratin that he had first observed with wool fibres. The conformational change in keratin gave a beautiful explanation of the elasticity of wool, but it would have been a very different proposition entirely to try and explain how the change from the 'A' to the 'B' form of DNA could account for the transmission of biological information. As it stood, wool, muscles, and flagella were therefore the immediate object of his attention.

Watson also had the advantage of having read a paper on diffraction theory published in 1951 by two crystallographers, William Cochran and Vladimir Vand, together with his Cambridge colleague Francis Crick. In this paper, Cochran, Crick, and Vand showed that X-ray diffraction of a helical molecule should give a pattern of spots in the form of a cross—a result that they acknowledged had also been shown a year earlier, but never published, by the mathematician Alec Stokes, a colleague of Maurice Wilkins at King's College.[63] Whether Astbury had ever read Cochrane, Crick, and Vand's paper is not clear, but the consensus seems to be that he had not kept abreast of these latest developments in the theory of X-ray diffraction.[64] Possibly this was due to the loss of heart that Davies and others described, and it raises the question of whether, had he been aware of these developments, he would have recognised a helical structure from the cross pattern in Beighton's photographs?

It is possible, but, in all likelihood, this alone would not have been enough to enable him to solve the structure of DNA, for, in addition to having read Cochrane, Crick, and Vand's paper, Watson was in possession of several extra pieces of information that were to prove vital. Of these, one had originated from the work done by the WIRA scientists at the Torridon facility in Leeds. By adapting their chromatography technique and applying it to sequencing of nucleic acids, Chargaff had not only demolished the tetranucleotide theory of DNA, but also showed that the base composition of nucleic acids varied between species, hinting that species-specific information might well be somehow carried within the base sequence. This in itself was a monumental contribution, but Chargaff's work had also yielded one other result that was to be of key importance for Watson and Crick.

At the end of 1949, having analysed the base composition of nucleic acids from a diverse range of organisms, including calf, yeast and tubercle bacillus, Chargaff noted that, in all cases, the bases always occurred in 'certain striking, but perhaps meaningless regularities'. [65] By 1950, he had explored these regularities a little more closely and noted that the ratios of adenine to thymine and of cytosine to guanine were always very close to one. Unlike Astbury, who, always excitable at any numerical correspondence, would have leapt upon this data and imbued it with some profound significance, Chargaff took a sober approach, pointing out that, while these values might just be mere accident, they might also speak of 'certain structural principles' that were fundamental to nucleic acids.[66]

Chargaff never appreciated at the time just how fundamental was his discovery, but others did. In 1952, while on a trip to Europe, he made a visit to Cambridge, where he was introduced to Watson and Crick. The duo did not make a good impression on Chargaff. Crick could not remember the exact names of the bases and was described by Chargaff as having 'the looks of a faded racing tout' who spoke in 'an incessant falsetto, with occasional nuggets glittering in the turbid stream of prattle'. [67] Watson, meanwhile, was 'quite undeveloped at twenty-three, a grin, more sly than sheepish; saying little, nothing of consequence'. [68]

The effect of Chargaff on Watson and Crick, by contrast, was 'electric'.[69] The reason for their excitement was that Chargaff's discovery about the ratios of bases in DNA could possibly explain another piece of work on the chemistry of nucleic acids by the Nottingham-based professor of chemistry John Mason Gulland. By performing careful electrometric titrations, Gulland had shown that nucleic acids contained a large number of electrostatic attractions known as hydrogen bonds that occurred between pairs of bases formed respectively by adenine and thymine and by guanine and cytosine.[70] This would explain perfectly why Chargaff had found that the ratio of these bases was always close to one.

But Gulland's work also contained another important clue. On addition of acid or alkali, he found that the viscosity of nucleic acids decreased dramatically. Gulland's explanation for this effect was that it occurred because these hydrogen bonds were formed between bases not on the same molecular chain, but on *different* molecular chains. The change in pH brought about by the addition of acid or alkali disrupted these inter-chain bonds, thus causing the different chains to separate and so make the solution less viscous.[71]

It was thanks to all this additional information that when Watson recognised from Franklin's 'Photo 51' that DNA must have a helical shape, his pulse began to race. For, with Chargaff's ratios of bases, Gulland's work on hydrogen bonding, and help Jerry Donohue, a chemist who showed Watson the specific geometry of these hydrogen bonds, and extra X-ray data from Rosalind Franklin showing that the DNA molecule consisted of two chains running in opposite directions, the helical shape took on a powerful significance.[72] As the two chains of nucleic acid twisted around each other, the bases on one chain formed specific pairs through hydrogen bonding with their opposite number on the other chain—always adenine on one chain pairing with thymine on the other, and similarly guanine with cytosine. It was this 'base-pairing', as it came to be known, that explained the fundamental property of DNA—that it could replicate to make copies of itself. During replication, both strands peeled apart, and each acted as a template for synthesis of a new strand. Thanks to the base-pairing rules, the base sequence of each newly synthesized strand was a complementary copy of the original template.

By bringing all of this information together, Watson and Crick were able to recognise the significance of the regular, well-defined structure that was suggested by the cross pattern in Franklin's 'Photo 51'. Had this same information been available to Astbury, might his pulse have started to race on seeing Beighton's pictures? He was certainly familiar with Gulland's research, to the extent that he had even sent Gulland samples of DNA for his titration work and shared correspondence with him. [73,74] But Astbury seems to have paid little attention to both Gulland's work on hydrogen bonding and his suggestion that this occurred between separate chains of nucleic acid. Maybe Astbury was distracted by the business of setting up his new department and perhaps in time he might have followed up Gulland's findings, but this will always remain mere speculation. Gulland's career was cut tragically short when he was killed in a train crash just south of Berwick in 1947. Had he survived, it is quite possible that Gulland might have been able to incorporate Astbury's X-ray work on DNA into his own work. For Astbury himself, there remained a far greater obstacle to understanding Beighton's patterns: his own success.[75]

Like Watson, Astbury had also taken note of Chargaff's work, but, whereas Watson's interest was in the specific ratios in which the bases occurred, Astbury was more interested in Chargaff's suggestion that the variation in base composition might give rise to geometric, structural variations that could be detectable by X-ray diffraction. This was the key difference between Watson and Astbury's interpretation of what the structure of DNA meant. For Watson, the helical shape of the molecule was the means by which DNA could copy itself and pass on biological information. A month after publishing their structure for DNA in April 1953, Watson and Crick published a second paper in *Nature* in which they proposed that the actual information itself was contained in the linear sequence of bases running up the inside of the helix—the helical shape of the molecule was merely the means by which this information was copied from one generation to the next. For Astbury, by contrast, it was the very shape of the DNA molecule itself that carried these biological properties through variations in its structure.

What he had hoped to find when Beighton analysed Chargaff's samples was a molecule that either showed variation in its structure or changed its configuration, but what he actually found was exactly the opposite. Instead of a long chain molecule the shape of which was highly variable, all Astbury would have seen was a monotonous repeating spiral pattern that held no promise of explaining how nucleic acids could carry biological traits, such as Avery had shown in pneumococcus.[76] Until now, this idea about the importance of shape and structural variation in fibres had provided a unifying principle with which Astbury had successfully explained a diverse range of phenomena, including how keratin fibres elongate to make wool elastic, how muscles contract, and how bacteria can swim. When Astbury had started working with Preston back in the 1930s, he was convinced that the importance of changes in molecular shape was not only confined to fibrous proteins like keratin, but also could be extended to include other types of biological fibre like the polysaccharides on the surface of cells such as *Valonia* and pneumococcus. With this one great idea that biological systems could be explained in terms of changes in the shape of their molecular chains, he felt that he had glimpsed a secret that promised to unify all of biology.[77] It meant, he said, 'something big, a new co-ordination of the activities of the proteins and the polysaccharides in one grand leap from an alga to a coccus'.[78] But, when applied to nucleic acids, it proved to be one grand leap too far. Franklin's 'Photo 51' may well have caused Watson great excitement, but Beighton's pictures would have been an utter disappointment, if not a source of outright frustration, for Astbury.[79]

Astbury's story is a cautionary tale for all those who believe that the task of science is to deliver ever more grand, all-encompassing 'theories of everything'. As Astbury's case shows, such aims are not without pitfalls and the search for unifying principles can even obscure valuable new insights, as happened in the case of Astbury and Beighton's photographs. Astbury's own response to Watson and Crick's discovery in 1953 is conspicuous only by its absence.[80] This might seem surprising at first—after all, the discovery of the double helix was surely a revolutionary moment in the history of science? But it is worth bearing in mind that revolutions, at least scientific ones, tend to be constructed with the benefit of hindsight. Today, rarely a week passes when DNA is not mentioned in the news and, thanks to the development of recombinant insulin, cloned animals, and stem-cell research, it is very easy for us at the start of the twenty-first century to look back and assume that April 1953 marked a momentous upheaval in science.

Actually, the truth is rather different. As science historian Robert Olby has shown, the publication of Watson and Crick's model did not usher in an overnight revolution and the response from the scientific community was rather muted to say the least,[81] or, as the science writer Matt Ridley put it:

> It was a momentous spring: Everest climbed, Elizabeth crowned, Stalin dead, Playboy born. The biggest event of all—life solved—caused barely a ripple.[82]

Ridley elegantly captures the mood, but while the claim that 'life had been solved' is a good rhetorical touch that emphasises the importance of Watson and Crick's work, it is far from the truth. Life had, most definitely, not been solved. The discovery of the double-helical structure of DNA raised as many questions as it answered: the molecule was two chains twisted around each other—very interesting, but so what? This might explain how the molecule copied itself, but many other questions remained unanswered and were not immediately obvious. How, for example, did the two chains physically unwind in order to make copies of themselves? And how did the sequence of bases running up each chain carry specific biological traits? And how did this information allow a single fertilised egg to develop into a complex multicellular organism? Slowly, over the next couple of decades, answers began to these questions began to emerge—along with further questions. Ultimately, there was no earth-shattering revolution when Watson and Crick published about the structure in 1953. Many scientists at the time felt that, although DNA was obviously central to biological processes, biology and genes could not simply be reduced to nucleic acids.[83]

Astbury was one such voice, and his understanding of what a gene was might also explain why he did nothing with Beighton's photographs. Today, we hear much talk about genes as if they were discrete, particulate entities in rather the same way as rocks, atoms, or stars. Our current view is that the linear sequence of amino acids in every protein in the cell is dictated by a corresponding well-defined sequence of bases at a particular region on a certain chromosome. We have come to call such precisely defined regions 'genes', and this has given rise to a language beloved of the popular press and media that every single human trait, no matter how complex, can be located in such segments of

DNA. Rarely a week passes without an article in a newspaper proclaiming the discovery of 'the gene for' yet another behavioural trait, be it criminality, violin playing, or preference for a particular soap opera.

But the idea of the gene over the course of the twentieth century was a malleable one. Astbury himself once asked 'What then is the gene? Is it a structure that is handed on? Or is it a set of physico-chemical conditions?'[84] In answering his own question, he entertained both alternatives—sometimes he talked about genes in the same way that we might do today—as discrete material entities. Certainly by the end of the 1950s, he seemed committed to the idea that, thanks to Watson and Crick's double-helical structure, questions of heredity and genetics could be reduced to DNA.[85]

Yet, at other times, he seemed to opt for the second answer to his question, emphasizing that the 'master key' to all questions in biology, such as the mechanism of heredity and the development of a multicellular organism from a single fertilised egg, lay in understanding 'some special type of collaboration between the proteins and the nucleic acids'.[86] By the time that he gave his 1950 Harvey Lecture 'Adventures in Molecular Biology', Astbury was still ascribing an almost-mystical significance to the near-identical correlation in the distance between adjacent bases in nucleic acids and adjacent amino acids in the protein chains of keratin that he and Bell had discovered in the late 1930s, saying that it could hardly just be 'a piece of pure numerology'.[87] In fact, Olby has argued that it was because Astbury was so seriously misled by his insistence that this correspondence must have some profound significance that it actually obstructed his work on nucleic acids.[88]

But Astbury's emphasis on the gene as an *interaction* between proteins and nucleic acids is revealing. To stress just how fundamental this interaction was to biology, he said that it was 'one of the great biological developments of our time'.[89] Rather than thinking of the gene as a solid, particulate unit of DNA, he maintained that reproduction and growth were controlled by:

'viable growth complexes'—a minimum specific combination of protein and nucleic acid that, given an adequate physic-chemical environment, can keep turning out the same molecular pattern indefinitely.[90]

For all that he championed the understanding of living systems in terms of their component molecules, he never felt that this was an end in itself, stressing that the 'result is an organisation, a structure with directional properties' through which 'we can always glimpse the proteins, not alone, but always working in collaboration, with sugars, phosphoric acid, bases, nucleic acid . . .'.[91] Understanding the components alone was not enough—they had to be understood in the context of the overall organisation and system of which they were part. For Astbury, both the carrier of the biological information and the agents of its execution were inseparable in understanding what he called 'the quest for the secret of biogenesis and the transmission of the patterns of life', and both were to be encompassed by the term 'gene'.[92]

Astbury's alternative concept of the gene as a set of interactions and processes rather than a material entity has some resonance with the views of certain contemporary biologists and philosophers of biology who argue that our current model, so beloved of the media, in which the gene is simply a defined, discrete sequence of DNA, is starting to look rather frayed through decades of over-use.[93] For example, biologist Steven Rose shares not only Astbury's taste for musical metaphor when describing living systems, but also his conviction that genes need to be understood as dynamic interactions with the wider context of the living cells of which they were part:

> Far from being isolated in the cell nucleus, magisterially issuing orders by which the rest of the cell is commanded, genes are in constant dynamic exchange with their cellular environment. The gene as a unit determinant of a character remains a convenient Mendelian abstraction, suitable for armchair theorists and computer-modellers with digital mind-sets. The gene as an active part in the cellular orchestra in any individual's lifeline is a very different proposition . . . The organism is both the weaver and the pattern it weaves, the choreographer and the dance that is danced. [94]

It may also help in explaining why Astbury did nothing with Beighton's pictures. In his memoir, Watson describes how, following their discovery, Crick allegedly went sprinting into 'The Eagle' pub in Cambridge declaring to anyone within earshot that he and Watson had discovered the secret of life. Although Crick himself later said that he had no recollection of this event and Watson has since admitted it that it was an act of poetic licence on his part, it nevertheless conveys dramatically Crick's sense that DNA was the key to understanding biology.[95,96,97] Yet even had he been able to solve the structure of DNA from Beighton's photographs, it is unlikely that Astbury would have gone sprinting into any one of the several pubs near the University of Leeds campus. For him, biology was bigger than simply DNA and with Beighton's photographs proving such a disappointment, Astbury chose not to publish them.

How very differently might history have unfolded had he done so— or even if he had simply made it known that he was making a new attempt to obtain X-ray pictures of DNA. During that same spring when Beighton was working on the new samples sent by Chargaff, Astbury was at the meeting in Naples where Wilkins presented his own X-ray diffraction photographs of DNA, which spurred Watson into action. Supposing that Astbury had casually mentioned to Watson that his research assistant back in Leeds was taking X-ray pictures of some new samples of DNA? On his arrival in Cambridge later that year, might Watson have been sufficiently curious to pay Astbury a visit in Leeds and be shown the black cross in Beighton's photographs, identical to that taken by Franklin and Gosling the following year? Perhaps he might even have scribbled Beighton's patterns down in the margin of his newspaper on the train from Leeds to Cambridge, ready to show them to Crick on his return.

It is certainly an intriguing alternative historical route by which Watson and Crick might have arrived at the double helix. From the point of view of the science involved, there seems to be no reason why events could not have unfolded in this way. But while the science presents no problem for this counterfactual history, the human element does. Wilkins may have been amused by Astbury and his 'little jokes' when he first met him at the Zoological Station, but Watson was not.[98] Astbury might have been a pioneer once, but for Watson he was a man who, with his best days long behind him, could now only make his presence felt by hovering around 'telling off-colour jokes'.[99] Disappointed by Astbury, Watson turned instead to Wilkins for help in solving the structure of DNA. Yet perhaps Watson was too hasty in his dismissal. For all that his best years might have been behind him, a group photograph taken at a conference on protein structure held in Pasadena in 1953 shows Astbury

Figure 23 Assembled delegates at the Conference on Protein Structure held in Pasadena, 1953. According to the archive website at Oregon State University (<http://osulibrary.oregonstate.edu/specialcollections/coll/ pauling/dna/ pictures/1953i.19-large.html>), Linus Pauling's listing of those pictured is as follows: Back row: Wilson, Perutz, Schomaker, Watson, Dunitz, Huxley, Crick, Marsh, Pasternak, Schroeder, Lindley, Trueblood, Bruchner, Huggins, Pepinsky, Palmer, Rollett, Riley, Luzzate, Freeman, Beadle, Davies, Tyler. Front row: Wilkins, Kendrew, Rich, Magdoff, King, Pauling, Krimm, Corey, Harker, Sutherland, Astbury, Bear, Bragg, Patterson, Edsall, Schmitt, MacArthur, Furnas, Randall, Elliott, Low, Itano, Trotter, Hughes. Photograph in Astbury papers, MS419 Box A.3, University of Leeds Special Collections, Brotherton Library. Reproduced by courtesy of the Archives, California Institute of Technology.

still standing proudly at the front of the assembled pioneers in molecular biology (Figure 23). Behind him, on the back row, are Watson and Crick, and what neither of them could have realised at the time was just what an important, if unwitting, part Astbury had played in their success earlier that year.

10

'The Road Not Taken . . .'[1]

Watson might have dismissed Astbury as a man whose best time was long past, but he and Crick were all too aware that they faced serious competition—not from Leeds, nor from King's College, London, but from Caltech in Pasadena. Following the success of his model of how protein chains could coil up into an alpha helix, Pauling was now beginning to show an interest in DNA. He had long been interested in how the genetic molecule, whether it be DNA or protein, could duplicate itself. In a 1948 lecture given in Nottingham, he had proposed that the key feature of such a molecule must be that it was somehow composed of two complementary parts, each of which could act as a template for replication of the other.[2] At that time however, Pauling was more inclined to the view that it was proteins, not DNA, that carried genetic information.[3]

By the early 1950s, however, while his attention was still firmly on protein structure, he was also beginning to take an interest in nucleic acids. In *The Double Helix*, Watson said that at first this prospect did not bother him, as he considered that Pauling's main focus was on protein structure.[4] But he nevertheless also admitted:

> We always worried, of course that maybe Linus Pauling would start thinking about it . . . But then in December, we heard, via his son Peter, who was then a student in the Cavendish, that his father had a structure for DNA. And that made us very, very curious.[5]

Anthony Serafini, one of Pauling's biographers, suggests that when Watson describes his response to Pauling's interest in DNA as being 'curious' he is actually being rather disingenuous—a more accurate word would be 'terrified'.[6] Similarly, Crick warned his boss Bragg that Pauling 'was far too dangerous to be allowed a second crack at DNA while the people on this side of the Atlantic sat on their hands'[7] and Pauling's own son Peter, who was doing his PhD at the Cavendish Laboratory in Cambridge, wrote to his father saying:

> You know how children are threatened 'You had better be good or the bad ogre will come get you.' Well, for more than a year, Francis and others have been saying to the nucleic acid people at King's, 'You had better work hard or Pauling will get interested in nucleic acids'.[8]

In 1951, Pauling had discussed the chemical composition of nucleic acids with the chemist Gerald Oster.[9] He knew the chemical details of the four bases that composed DNA, but what intrigued him was how these bases were arranged in space to form a giant nucleic acid chain. Since his success in solving the helical structure of protein chains had begun from first principles about the size of their constituent amino acids and the specific way in which they form chemical bonds with each other, he reasoned that the same approach should be equally fruitful when applied to DNA. But his success in solving the alpha helix had not been built on theory and model building alone—it had involved the meticulous process of taking thousands of X-ray photographs, and if his work on DNA was to be equally successful, it too would require structural data in the form of good-quality X-ray diffraction photographs.

He was delighted therefore to learn from Oster that Maurice Wilkins had obtained the excellent photographs of crystalline 'A'-form DNA that were presented at the conference in Naples in the spring of 1951 and that proved to be such an inspiration to Watson. While Pauling had not yet seen the pictures himself, mere knowledge of their existence had a similar effect in spurring him into action, and he wrote to Wilkins asking whether it might be possible to see them. But Wilkins was reluctant to share them. All too aware that Pauling had beaten Bragg at Cambridge to solving the helical structure of protein chains, Wilkins suspected that Pauling would easily repeat this success with nucleic acids if he had access to decent X-ray photographs. After pondering his response for a week, he finally replied saying that he needed more time to interpret his own pictures.

Pauling was persistent. Later in the year, he wrote a second letter to King's, but this time addressed it to J. T. Randall, Wilkins' boss; again this met with no success. Randall was blunt in his reply, saying: 'Wilkins and others are busily engaged in working out the interpretation of the desoxy-ribosenucleic acid x-ray photographs, and it would not be fair to them, or to the efforts of our laboratory as a whole, to hand these over to you.'[10,11]

Unlike Watson and Crick or Franklin and Wilkins, Pauling lacked the luxury of being able to give DNA his undivided attention. His distraction from the problem was not without good reason. His model of the coiling of a protein chain into an alpha helix was by no means universally accepted and was under attack from the giants of protein structure on the other side of the Atlantic: Bernal, Bragg, and Astbury. In 1951, the Royal Society began organising a conference on protein structure to be held the following year. The British protein scientists hoped that the conference would be a 'high noon' for Pauling's alpha helix, and in their vanguard was Astbury, who remained unconvinced by Pauling's arguments that proteins could form helical chains. Despite having enjoyed a cordial relationship and correspondence since their first meeting back in 1937, Astbury's invitation to Pauling asking that he come and defend his ideas at the meeting so that 'we really could have a first-class row' was perhaps said only half in jest.[12]

On 5 April 1952, Astbury wrote to Pauling informing him that he had booked a room for him to stay at the Hotel Rembrandt in South Kensington from 28 April to 5 May, during which time the conference would be held.[13] A couple of weeks later, Astbury received a reply from Pauling thanking him for his help but expressing his disappointment that he would now be unable to attend the meeting as it had been decided that 'my proposed travel would not be in the best interests of the USA'.[14] Pauling was facing far bigger distractions than any challenge to his scientific ideas. He was now famous not only for his science but also for his politics. Since giving his first political speech in 1940 arguing for the need to resist fascism, Pauling had become increasingly active in causes close to the heart of the political left. After the dropping of the atomic bombs on Hiroshima and Nagasaki, he had become ever more vocal in his concerns about nuclear weapons and the possibility of a deadly arms race with the Soviet Union. With politicians in the USA becoming seriously worried about the threat of communist agents operating on the domestic front, such an overt declaration ran the risk of attracting suspicion. As this concern gathered momentum, it became an active crusade led by Senator Joseph McCarthy and the House Un-American Activities Committee (HUAC) to hunt out 'Reds under the Bed'.

The paranoia was rooted in an anxiety that the Cold War might well turn hot at any moment—a fear that seemed quite reasonable when the Soviet Union first tested its own nuclear weapon in 1949 and communist North Korea invaded its southern neighbour in June 1950. It was against this background of fear that Pauling spoke out against measures such as the loyalty oaths that the University of California had recently imposed upon employees, and he called for negotiations with the Soviet Union and for National Science Foundation funding into the causes and prevention of war. All the time, he was actively monitored by the FBI and denounced by HUAC as being one of the foremost leaders in a 'Campaign to Disarm and Defeat the United States'.[15]

Ironically, at the same time that he was under suspicion of being a communist spy by the security services at home, Pauling was also being denounced in the Soviet Union. Writing in both *Pravda* and *Nature*, several Soviet scientists attacked Pauling's theory of chemical bonding between atoms on the grounds that it contradicted the philosophical materialism at the heart of Marxist–Leninist orthodoxy.[16,17] But this won him no favours with HUAC and the McCarthyites, and in 1952 Pauling was informed by the US State Department that his passport had been cancelled.[18,19] Under suspicion of being a communist, he was not permitted to leave the USA and was therefore unable to attend the Royal Society meeting in London in May 1952. On hearing that Pauling's passport had been cancelled, the organisers of the meeting and the wider international scientific community were outraged at his treatment. Astbury wrote a letter of support to Pauling, saying:

> I had already written to you to say how disappointed, and indeed indignant, we were to learn this at the last moment, and I must say that reading your documents only intensifies these feelings. I am no communist, and in fact I find most manifestations of Communism loathsome and dishonest—witness the utterly fantastic Soviet attack recently reported in *Nature* on your own resonance theory You are a scientist who belongs to the world, if I may say so, and in Britain particularly you are an honoured guest everywhere. I can proclaim this all the more sincerely because, frankly I disagree strongly with some of your conclusions on protein structure—all the more reason for you to come over here and let us get to grips with you, so to speak.[20]

Albert Einstein was another stalwart defender of Pauling and wrote to the US Secretary of State, saying:

Professor Pauling is one of the most prominent and inventive scientists in this country. I have the highest esteem for his character and for his reliability as a man and as a citizen. To make it impossible for him by governmental action to travel abroad would—according to my conviction—be seriously detrimental to the interest and reputation of this country.[21]

The US State Department could never have anticipated just how seriously detrimental the cancellation of Pauling's passport would turn out to be for US interests. All of Pauling's biographers agree that this was a severe blow to his work on DNA, and Serafini suggests that, had Pauling not become a pawn in Cold War politics, events leading up to the discovery of the structure of DNA might well have unfolded very differently:

One result of the denial of Pauling's passport extends far beyond other matters in historical importance. Had he been able to travel to England in 1952, he might well have beaten Watson and Crick in the legendary and much publicized race to unravel the structure of the DNA molecule.[22]

The consensus of Pauling's biographers is that making the trip to the Royal Society meeting in May 1952 would have given Pauling the key data that he so desperately needed to solve the structure of DNA: access to good quality X-ray photographs like those taken by Wilkins. It is a view shared by the historian Robert Olby, who suggests that Pauling 'would very likely have been able to see the King's pictures that year had the U.S. government not refused him a passport, thus preventing him from attending the one-day Royal Society protein meeting in London'.[23],[24]

What Pauling would not have seen, however, even had he made the trip, was Franklin's 'Photo 51'. Before going along to attend the Royal Society meeting, Franklin had set up her X-ray camera to take pictures of some newly prepared fibres of DNA. Having left the equipment to run while she was at the meeting, on her return she developed the films to find that they showed a striking pattern of black spots in the shape of a cross. She knew that this meant she was looking at the 'B' form of DNA, in which the molecule contained much more water than the crystalline 'A' form on which she had done most of her work. It was the 'A' form, however, that gave far more spots on a photographic film, and, to a crystallographer, more spots meant more information about the molecule. Labelling the picture 'Photo 51', she filed it in her laboratory notes, intending to return to it when she had extracted as much information as she could from her pictures of the 'A' form.

Yet even without seeing the distinctive cross pattern of 'Photo 51', the other photographs of 'A'-form DNA already taken by Wilkins and Franklin would have contained enough information about distances within the molecule to force Pauling to seriously rethink his hopelessly flawed structure for DNA. Had he ever been able to see these pictures, they alone could well have been enough to give him a head start over Watson and Crick. As it happened, his associate Robert Corey, with whom he had published his model of the alpha helix in proteins, was present in London and was shown some pictures of DNA by Franklin. Corey told Pauling that they were of high quality, but this was a poor substitute for Pauling seeing the pictures for himself and, in the end, it was simply not enough.

There was also the added problem of the demand that Pauling's need to defend his alpha-helical model of protein structure was placing on his time and intellectual energy. Following the outcry by the international scientific community at the cancellation of Pauling's passport, the US State Department reconsidered its decision and finally allowed Pauling to travel outside the USA under very strict limitations. Under these conditions, Pauling travelled to Paris in July 1952, where he was to be honorary chairman of the International Congress of Biochemistry. It was while on this trip to France that he first heard news of Hershey and Chase's experiments on radiolabelled bacteriophages that showed that only the nucleic acid of the phage entered an infected bacterium and so provided strong evidence in support of Avery's work eight years earlier on pneumococcus.

The Hershey–Chase experiment certainly suggested to Pauling that solving the structure of DNA was a problem well worth solving, but his more immediate concern was to defend his helical model of protein chains against its critics—for the moment, DNA could wait.

It was to discuss the helix with its detractors such as Bernal, Bragg, and Astbury that Pauling left France and crossed the channel to England. Here he toured the major centres of protein structure research, including making the trip up to Leeds, where he had an enjoyable stay at Astbury's house. What he did not do on this trip, and which may with hindsight seem rather surprising, was visit the laboratory at King's to see Wilkins and Franklin and their X-ray pictures of DNA. It seems that concerns about protein structure and fending off the criticisms of his alpha-helical model of polypeptide chains had become so dominant in his mind that he simply could not spare the time to be too distracted by DNA.[25]

Nevertheless, Pauling's interest in nucleic acids gained new momentum once he was back in the USA. In November 1952, he attended a seminar at Caltech given by visiting biologist Robley Williams, who was working on the structure of the tobacco mosaic virus. Williams showed some data suggesting that the ribonucleic acid (RNA) chains in the virus might be coiled into helices. This prompted Pauling to start reflecting once again on his own ideas about the structure of nucleic acids. The same method that had proved so successful in solving the helical structure of protein chains could surely be applied to DNA: start with knowledge of the chemistry and size of the individual components of the chain, then use this information to build a theoretical structural model based on information from X-ray photographs.

Yet without direct access to Wilkins' or Franklin and Gosling's X-ray photographs, Pauling had to fall back on using the only others available: those taken by Astbury with Bell in 1938 and also some new pictures taken in 1947. But what Pauling did not realise, like Astbury, was that these were actually a mix of both 'A' and 'B' forms, which meant that they gave ambiguous and limited structural information. Using these pictures, Pauling began making the calculations for a hypothetical model of DNA.

His confidence that this approach would bring him the same success as it had done in solving the alpha helix was, however, severely misplaced. From Astbury's pictures and measurements of the density of DNA, Pauling concluded that each DNA molecule must be composed of three nucleic acid chains, not two as in Watson and Crick's model. The problem was also that Astbury's measurements of the density of DNA suggested that it was a very tightly packed molecule. Pauling reasoned that, as the four bases were very different sizes, it would be impossible for them to pack tightly together on the inside of the DNA chains, and so they must therefore run up the *outside* of the nucleic acid chains, jutting out into the surrounding solution, while the phosphates were packed tightly on the inside. It was the complete opposite of Watson and Crick's model, and hopelessly wrong.

On the last day of December 1952, Pauling wrote a letter to Randall at King's, telling him:

> Professor Corey and I are especially happy during this holiday season. We have been attacking the problem of the structure of nucleic acids during recent months and . . . we feel that the nucleic acid molecule may have one and only one stable structure.[26]

On that same day, he and Corey submitted their paper 'A proposed structure for the nucleic acids' to the *Proceedings of the National Academy of Sciences* and the paper was accepted for publication in February of the following year.[27] It was when Watson received a manuscript of this paper that he quickly jumped on the train to London to consult Wilkins and Franklin, fearing that Pauling might be about to snatch the prize of DNA from him.

He need not have worried. Unlike Watson and Crick, Pauling and Corey's model gave no hint at the most important question to be answered—how the DNA molecule could replicate itself. Moreover, the flaws in Pauling's model were quickly apparent to everyone who looked at it, leaving them rather baffled as to how a structural chemist of Pauling's stature had not spotted them himself. In Watson's opinion, any student who made similar errors would not have been allowed to study at Caltech.

Why then did Pauling and Corey rush into print with such a poorly considered model? For one of his biographers, the reasons for Pauling's failure were twofold and simple—'hurry and hubris'.[28] The reason for Pauling's hurry was that he knew of the work that was going on at both Cambridge and London and feared that one of these groups would soon solve the structure of DNA. In response, he felt the need to establish a priority claim on the territory of nucleic acid structure by submitting a paper quickly. This might explain why Pauling and Corey never actually stated in their paper that this structure had been subjected to rigorous experimental test—they were quite clear that it was a tentative hypothetical model.

The hubris lay in his confidence that his success in solving the helical structure of keratin chains could be easily repeated with DNA. The success of the alpha helix had been both a blessing and a curse to Pauling—a blessing in that it brought him a Nobel Prize in 1954, a curse in that it left him thinking that the structure of DNA would simply fall into his lap without needing 'to do the homework required by others'.[29]

Speaking many years later, Pauling said that when his wife had asked him why he hadn't simply worked harder at the problem of DNA, he had replied: 'I guess that I always thought that the DNA structure was mine to solve, and therefore I didn't pursue it aggressively enough.'[30]

Even though word began to spread that Pauling's structure was wrong, he still held out a faint hope that he might be proved right. He wrote to Watson saying that he hoped that data from Franklin and Wilkins at King's might well decide which of the two structures was right. Of course, what Pauling did not realise is that Watson and Crick had already seen Franklin's 'Photo 51'. They had the one clue that he had never been able to obtain—a high-quality X-ray photograph of 'B'-form DNA. For Ridley, it was Pauling, not Franklin, who was 'the one most cheated of destiny by Watson's haste'.[31]

How very differently events might have unfolded if Pauling had had access to good-quality X-ray pictures of the 'B' form of DNA. In his letter to Randall written on 31 December 1952, the same day that he submitted his paper, he underlined the importance of X-ray data, saying:

> I regret to say that our X-ray photographs of sodium thymonucleate are not especially good; I have never seen the photographs made in your laboratory [*those of Wilkins and Franklin*] but I understand that they are much better than those of Astbury and Bell, whereas ours are inferior to Astbury and Bell's. We are hoping to obtain better photographs, but fortunately the photographs that we have are good enough to permit the derivation of our structure.[32]

Pauling could not have been more wrong in placing such faith in the quality of the X-ray data to support his proposed model for DNA. In summarising the reasons why Pauling never solved the structure of DNA, biographer Thomas Hager suggests that:

> the real problem was not the passport policy. Instead, three unrelated factors combined to set Pauling wrong. The first was his focus on proteins to the exclusion of almost everything else. The second was inadequate data. The X-ray photos he was using were taken of a mixture of two forms of DNA and were almost worthless. The third was pride. He simply did not feel that he needed to pursue DNA full tilt.[33]

All three of the problems listed by Hager would have been solved by access to a good-quality picture like Franklin's 'Photo 51'. Such a picture would not only have given Pauling the structural data that he needed to build an accurate model, but also have shaken him out of the complacency that resulted from his success with the alpha helix and made him less prone to be distracted by problems of protein structure. And an opportunity to see just such a picture could so easily have arisen—not when visiting London during the summer of 1952, but a year earlier when Astbury's research assistant Beighton had taken his beautiful photographs showing the cross-pattern of 'B'-form DNA.

Had Astbury published these pictures in 1951 when Pauling was riding the crest of his triumph in predicting the helical coiling of protein chains, it is quite likely that Pauling would have seized them in an instant. Thus armed, he would have had a two-year start over Watson and Crick, and, in all likelihood, it would be Caltech that would today be celebrating the discovery of the double-helical structure of DNA and not Cambridge. Pride, cancelled passports, and distraction with proteins may all well have contributed to Pauling's failure to solve DNA, but perhaps the biggest and to date most overlooked reason lies with Astbury and Beighton's forgotten photographs.

Even if Astbury did not publish Beighton's photographs of 'B'-form DNA, is it possible that he might have mentioned them to Pauling when he came up to Leeds in late summer 1952? On reflection, it is unlikely, for Astbury's attention at the time was elsewhere. Pauling's model for the alpha-helical coiling of polypeptide chains dominated the field of protein research at that time, and it was this, not nucleic acids, that would have been at the forefront of both men's minds. And, in addition, Astbury was in the throes of delight over the X-ray diffraction studies of bacterial flagella.

When Pauling stayed at Astbury's house in Headingley, little could he have realised that while his host showed him diagrams of flagella and how they might work, he also had in his possession the one clue that Pauling needed to solve the structure of DNA.[34]

Had Astbury revealed Beighton's photograph to Pauling, as Wilkins had done with Franklin's photograph to Watson, then history might well have been made that summer at 189 Kirkstall Lane directly over the road from Headingley Cricket Ground. Instead, it is for cricketing triumphs that Headingley is remembered rather than as the site of a historical encounter between two giants of molecular biology.

Like the strands of the double helix, Astbury and Pauling's scientific paths and destinies were intricately entwined. It was Astbury's work on the structure of keratin that had first inspired Pauling to move from the study of inorganic crystals into biological molecules; it was the anomalous spot at 5.1 angstroms on Astbury's diffraction patterns of keratin that had misled Bragg and allowed Pauling to win the race to solve the alpha helix; now their individual failure to solve the structure of DNA also shared a certain symmetry. In both cases, their failure to solve DNA was rooted in their earlier scientific success. For Pauling, the success of the methods that he used in solving the alpha helix brought a certain degree of complacency when he turned to DNA. As he said himself, 'I always thought that sooner or later I would find the structure of DNA ... It was just a matter of time.'[35] For Astbury, meanwhile, it was the success of his early work on keratin chains in wool fibres that left him convinced that the key to understanding biological macromolecules lay in observing the variation and changes in their shape. It was a paradigm that explained beautifully how wool fibres stretched, how muscles contracted, and what happens when eggs are boiled, but it could give no account of how the monotonous, repeating spiral of DNA carried biological information.

But this was by no means the final twist as Pauling and Astbury's scientific destinies wound their way around each other. The latter part of Astbury's life was characterised not only by a disillusionment with certain aspects of science, but also by physical poor health. In 1955, he was invited to visit Moscow along with the British biologist Sir Peter Medawar and the quantum physicist Paul Dirac, but had to cancel the trip owing to ill health.[36] Later in that same year, he flew to Australia to attend a major international conference on wool, but on his return developed severe problems with blood clots that required him to spend six weeks in the Leeds General Infirmary (LGI).[37] During this time, Astbury was placed in the care of a recently qualified junior doctor. In the course of their daily conversations, it emerged that the young man was considering academic medical research, a path on which Astbury was keen to encourage him. When the young doctor arrived one day armed with a syringe, Astbury point blank refused to allow him to perform the injection unless the doctor was able to explain the exact effect that the administered substance would have on his blood chemistry. The young man duly hurried away to the library and, having consulted a few books and journals, was able on his return to give Astbury a full account in molecular terms of what the injection would do to him. Suitably impressed, Astbury immediately allowed him to administer the drug. This was not a case of a stubborn patient being obstructive. Rather, Astbury

was so impressed by the young man's enthusiasm for basic scientific research in medicine that he wanted to encourage him by instilling within
him the idea that living systems needed to be understood at the molecular level. It was to prove an instructive lesson that the young doctor
would never forget.[38]

His period in the LGI was instructive for Astbury too. On his recov-
ery, he wrote to the editor of the *Yorkshire Evening Post* praising the staff
at the LGI, saying that he had 'benefited physically, but that is not everything, it has been a grand spiritual experience too'.[39] Sadly, however,
this marked the beginning of a steady decline, and six years later, on 4
June 1961, he died having suffered a pulmonary infarct, auricular fib-
rillation, and mitral stenosis.[40] It is a testament to his character that,
despite this steady deterioration in his health, he never allowed his condition to diminish his *joie de vivre*, and the very night before his death had
been at a party bellowing in his jovial manner for his colleague Keith
Parker to bring him another whisky.[41]

In the eulogies and tributes that followed his death, he was hailed
as one of the pioneers of molecular biology, and it is perhaps for this
that he should be remembered, rather than the man who failed to
solve the structure of DNA despite coming so close. For, as Preston said
many years later: 'who could expect the man who cleared the forest
to erect in the clearing a nuclear power station?'[42] To focus on the fact
that he never published Beighton's pictures does him something of a
disservice, for his scientific legacy extended far beyond DNA. Preston
said of Astbury that 'He left behind him no formal school and few dis-
ciples. But he left as a monument the whole of molecular biology as a
testament to what he had been striving for'; following Astbury's death
it was Preston who continued Astbury's legacy at Leeds by becoming
Head of the new Astbury Department of Biophysics, which still exists
today as the Astbury Centre for Structural Molecular Biology and continues to be a centre of world-class research in the field of biomolecular
structure.[43]

But, beyond Leeds, it was Pauling in the USA who truly took up
the torch handed on by Astbury, for while Astbury's failure to pub-
licise Beighton's photographs may have denied Pauling the structure
of DNA, it was through Pauling's subsequent work that the impact of
Astbury's legacy was truly felt.

Inspired by Astbury's work on the changes in folding of the keratin
chain, Pauling had become interested in haemoglobin, the protein that

transports oxygen from the lungs to the tissues, where the oxygen is released to generate energy by respiration. In the spring of 1945, Pauling was at a dinner meeting with some clinicians when the talk turned to a condition called sickle cell anaemia.[44] In this condition, the shape of the red blood cells becomes deformed from a flattened disc to a crescent, or sickle. The sickle-shaped cells then block up small blood vessels, ultimately giving rise to severe organ damage and a host of clinical complications due to poor blood supply and the impaired ability of the blood to carry oxygen.

One particular aspect of the disease intrigued Pauling. This was that the deformed sickle-shaped cells were far more commonly found in venous blood, where there was less oxygen. If the change in oxygen levels was causing the deformation, then Pauling reasoned that the protein haemoglobin must somehow be involved. He speculated that there must be some fundamental change in the structure of the polypeptide chain of haemoglobin that caused it to stick to other haemoglobin molecules and form insoluble aggregates. He proposed that it was this aggregation of the haemoglobin proteins that was causing the deformation of the red blood cell. Together with Harvey Itano, a physician, and John Singer, a post-doctoral fellow, Pauling began to investigate how changes in the haemoglobin chain might cause these damaging effects.

Pauling, Singer, and Itano showed that there was one crucial difference between the haemoglobin of a healthy person and that of someone with sickle-cell anaemia. The key method in discovering this difference was the electrophoresis method recently developed by Swedish chemist Arne Tiselius. When subjected to an electrical field, protein molecules with a net electrical charge would move towards either the positive or the negative end of the electrical gradient at a speed determined by their size, with smaller molecules moving faster. Using this method, Pauling, Singer, and Itano noticed that the sickle-cell haemoglobin moved toward the negative electrode faster than haemoglobin from a healthy patient, suggesting that it carried some extra positive charge. The implication was monumental: a slight difference in the electrical charge of the haemoglobin chains could cause them to misfold at low partial pressures of oxygen in such a way that they formed insoluble aggregates giving rise to the symptoms of sickle-cell anaemia.

In 1949 Pauling, Singer, and Itano published these findings in the journal *Science* under the title 'Sickle cell anaemia, a molecular disease'.[45] Their paper has since come to be recognised as a landmark publication in the history of medicine. Inspired by the work of Pauling's team, Max Perutz, who was making X-ray studies of normal haemoglobin at Cambridge, turned his attention to the sickle-cell form of the protein and provided experimental confirmation of Pauling's hypothesis that it became insoluble in the absence of oxygen. Biochemist Vernon Ingram, one of Perutz's team at Cambridge, then showed that the difference in the electrostatic charge of the molecule that Pauling had proposed was responsible for the misfolding of the sickle-cell protein arose from a single point mutation in which one amino acid of the polypeptide chain was substituted for another. Together with German doctor Hermann Lehmann, Perutz went on to classify the abnormal and clinical symptoms of 34 variants of haemoglobin, as well as speculating about the possibility of using drugs to prevent the aggregation of the mutated protein.[46]

The work on sickle-cell haemoglobin has since been hailed as 'the curtain-raiser for the era of molecular medicine'.[47] Thanks to Pauling, Perutz, and their associates, a pathological condition had been described for the first time in terms of molecular shape changes. At the core of this explanation was an idea that was rooted in Astbury's first work on the keratin protein chain. Pauling, Singer, and Itano's work on sickle-cell haemoglobin, followed by that of Perutz, marked the beginning of a new era in medicine, whereby the tools of molecular biology began to be used to understand disease at the molecular level. The understanding of pathology in terms of altered molecular shapes is central to one particular area of molecular medicine: the study of debilitating conditions such as Alzheimer's disease, Parkinson's disease, and prion diseases such as Creutzfeldt–Jakob disease (CJD). Known collectively as protein misfolding diseases, these are pathological conditions in which the polypeptide chains of key proteins in the brain such as beta-amyloid or the tau protein become refolded from a soluble form into insoluble fibres. As a result, the proteins precipitate out of solution in a fibrous form that causes damage to the neurons and ultimately gives rise to the clinical symptoms.[48] Our understanding of how these proteins change their shape from a soluble to the insoluble form is rooted in Astbury's observations that globular, soluble proteins such as egg albumin and the seed protein edestin could be refolded by chemical treatment into an insoluble fibrous form, an insight for which he is still given credit by name in contemporary research papers in this field.[49]

Understanding diseases such as sickle-cell anaemia in terms of a change in molecular shape was a powerful insight, but Astbury's ultimate legacy went still further. So far, molecular biology had been a science based purely on the observation of molecular structure. Using X-rays, Astbury, Bernal, and Pauling had been able to peer deeper and deeper into living systems until they were granted a glimpse of the giant molecular chains from which organisms were constructed. This vision offered by X-ray diffraction had enabled a radical new understanding of life at the molecular level—but why stop there? For, having understood the molecules of life, was not the next step surely to alter them for our own ends?

11

The Man in the Monkeynut Coat

In the mid-1970s, the *Boston Globe* newspaper featured a cartoon showing that stalwart of Hollywood 'B' sci-fi films, the mad scientist, resplendent with his wide staring eyes and wild hair racing into his laboratory to be greeted by hordes of waiting mutant creatures. In his hand, he is triumphantly waving a copy of the *Boston Globe* bearing the headline 'Cambridge Okays Genetic Research' and, in the caption beneath the cartoon, he declares with delight to his monstrous creations: 'Crack out the liquid nitrogen, dumplings . . . we're on our way'.[1]

The cartoon was just one of many that caught the mood of the time and gives an insight into fears that were circulating in the wider media about certain recent developments in the newly emerging science of recombinant DNA technology, or, as it was more popularly known, 'genetic engineering'. The source of the cartoon scientist's delight was the news that the town council of Cambridge, Massachusetts had just overturned a decision made by the town mayor to block plans by Harvard University to build a new research laboratory for the purpose of conducting experiments into recombinant DNA.[2]

At the heart of this new and controversial technology was the discovery made towards the end of the 1960s of a type of bacterial enzyme that could physically cut both strands of the double helix at specific points on the DNA molecule.[3,4] Having evolved as a primitive defence mechanism to limit or 'restrict' the infection of bacteria by viruses called bacteriophages, these enzymes were dubbed 'restriction' enzymes and they were to provide scientists with a powerful new tool.

Described by the young daughter of one of their discoverers as a 'servant with the scissors', restriction enzymes could cut DNA into fragments of a defined length and sequence, thus enabling the direct and deliberate manipulation of chains of nucleic acid.[5] Two pieces of DNA from completely unrelated organisms could now be excised at specific points and physically joined together to form a new, hybrid, or 'recombinant' DNA molecule. In 1972, this became a reality when Paul Berg, Professor of Biochemistry at Stanford University, used restriction enzymes to snip the specific sequence of DNA responsible for metabolism of the sugar galactose in the common bacterium *Escherichia coli* out of the bacterial chromosome and insert it into the genetic material of a simian virus that could infect human cells.[6] Berg showed that, by hitching a lift aboard the genome of a virus, bacterial DNA could be transported into human cells and that, once inside those human cells, the bacterial DNA remained active by producing bacterial proteins.

This technology raised fascinating new possibilities for the study of medicine and disease—but it also raised alarm. What if, for example, a sequence of DNA known to cause cancer in humans was introduced into the genome of a common bacterium such as *E. coli* that lives in the human gut? Or if DNA conferring powerful antibiotic resistance was introduced into a dangerous disease-causing bacterium, which then escaped from the laboratory?[7] Realising these potential dangers, Berg and 11 other leading researchers published a joint letter calling for a moratorium on recombinant DNA research until a set of guidelines for such work had been agreed upon.[8] As a result of Berg's letter, nearly 150 scientists from around the world gathered the following year at the old wooden chapel of the Asilomar Conference Centre on the Monterey Peninsula in Northern California to engage in a sometimes heated and chaotic debate about how recombinant DNA work should be regulated.[9,10] Following the Asilomar meeting, the National Institutes for Health (NIH) issued a set of rules in January 1976 for the conduct of experiments involving recombinant DNA.

But the NIH rules did little to allay the fears of the popular press, and a frenzied, emotive, and sometimes alarmist debate began in the wider media.[11] Speaking of recombinant DNA technology, the newspaper columnist Charles McCabe said 'it scares the daylights out of me. Jiggling with genes may cure cancer. Then again, it may cause outbreaks of new forms of cancer.'[12] Meanwhile, the President of the National Academy of Science in Washington DC received a letter asking whether alleged sightings of a 'strange orange-eyed creature' might have anything to do with recombinant DNA experiments taking place in the New England area.[13] It was amid this climate of fear and suspicion that the University of Harvard's plans for a new laboratory were initially opposed before finally being granted permission.

By the end of the decade, however, fears about being over-run by hordes of genetic hybrid monsters were receding as it became evident that molecular biology might also have the potential to generate enormous wealth. In 1976, Herbert Boyer, Professor of Biochemistry at University of California, San Francisco, together with the venture capitalist Robert Swanson, set up the world's first biotechnology company, Genentech, and began using recombinant DNA technology to produce human insulin for the treatment of diabetes.[14]

In October 1980, Boyer appeared on the front cover of *Time* magazine as Genentech was floated on the US stock market in what the *Wall Street Journal* described as 'One of the most spectacular debuts in market history'.[15] In a day of frenzied trading, during which over half the company's shares were resold by their purchasers, Genentech's stock witnessed a meteoric rise in price from $31 to $89, falling to $71 per share by the close of business. With over seven million shares now giving the company a value of $500 million, Swanson and Boyer very quickly became multimillionaires. Wall Street was hungry for molecular biology and, by 1983, over 200 small biotechnology companies had been founded in the USA. Speaking at a conference organised by the journal *Nature* that same year, Watson made the pronouncement that 'If you are young, there really is no option but to be a molecular biologist.'[16]

The kind of molecular biologists to which Watson was referring were no longer engaged merely in observing and solving the structure of large biological molecules. They were now a kind of molecular engineer—specialists in the isolation and deliberate manipulation of nucleic acid chains. Nor was this active manipulation of nucleic acid chains merely a means to answer questions about biology, for, as the scientists at Genentech had demonstrated, it could be a powerful tool for industry and be applied to very practical ends.

Molecular biology had undergone a profound transformation. It was no longer concerned with simply observing living systems at the molecular level, instead it was now about their active and deliberate alteration—whether for academic research or for industrial ends. Thanks largely to the spectacular success of Genentech and the numerous other small biotechnology companies that sprang up at around the same time, the origins of this transformation are usually associated with developments on the US West Coast in the mid to late 1970s. But a case can also be made that it had an alternative origin—one that occurred not under the sunny blue skies of 1970s California, but under the rather more leaden, rainy skies of West Yorkshire over three decades earlier.[17]

About three decades before the *Boston Globe* was running cartoons showing mad scientists and genetic monsters, the *Yorkshire Evening Post*, a provincial UK paper printed in Leeds, featured a small cartoon showing a gentleman wearing an overcoat on which several little birds had landed and were pecking at the fabric of the garment, much to the surprise of a passing lady and police officer (Figure 24).[18] The caption beneath the cartoon read:

> In a broadcast, a scientist said he had an overcoat made largely of peanuts.

The cartoon related to a series of radio lectures that Astbury had delivered on the BBC between 1942 and 1944.[19] The first of these was in a series called 'Science Lifts the Veil' and Astbury began by explaining that he had struggled to think of a title as his lecture encompassed such a variety of subjects:

> many things you might think had little or nothing to do with one another—things like wool and whale bone, toothbrushes and gun turrets, hair and muscle, leather and jellyfish, virus diseases and silk stockings and, I might even say, cabbages and kings.[20]

The diversity of natural forms was deceptive: beneath it all there lay the common unifying theme of the chain molecule. With these threadlike molecules Astbury said that Nature had bestowed:

> cunning properties on hair, nails and horn, muscles, tendons and skin, nerves, brain and bone, feathers, beaks and claws, rubber, plant juices and viruses, the skeletons of plants and insects and hosts of lowly creatures in the earth and in the waters under the earth. Above all other molecules, the thread molecule is the chosen instrument in the symphony of creation and we may well thrill to thoughts of these things as we thrill to mighty music.[21]

Understanding the shape of these molecules explained why wool stretched, how muscles contracted, and how bacteria could swim, but it also promised much more. Astbury proclaimed that unravelling the structure of these molecules would 'lead straight to the heart of the mystery of life' and, in so doing, grant us a 'glimpse of a truly splendid body of knowledge wherewith to help build our world anew'.[22]

In a broadcast, a scientist said
he had an overcoat made largely
of pea-nuts.

Figure 24 Cartoon from the Yorkshire Evening Post, 18 January 1944, with
the caption: *In a broadcast, a scientist said he had an overcoat made largely of pea-nuts.* Artist
unknown.

Astbury Papers, MS419 Box A.1 Press Cuttings, University of Leeds Special Collections.
Reproduced with the kind permission of the University of Leeds, Brotherton Library
Special Collections.

As he closed his lecture, Astbury suggested that an important part
of building the world anew would be the rise of novel industries based
not simply on understanding the structure of these giant molecules,
but on their deliberate manipulation and, to underline his point, he
drew the attention of his audience to the overcoat that he was wearing
(Figure 25).

(a)   (b)

Figure 25 Astbury's famous 'monkeynut overcoat'. Male model unknown. Astbury's daughter Maureen is shown sporting her 'monkeynut jumper'.

Photographs taken by Earle R. M. Brooke, 1948. Astbury Papers, MS419 Box A.2, University of Leeds Special Collections, Brotherton Library. Reproduced with the kind permission of the University of Leeds, Brotherton Library Special Collections.

At first sight, it was a rather ordinary overcoat as might be purchased in any gentleman's outfitters at the time, yet the fibres from which it had been spun were not wool but from a much more unusual source. In his lecture, Astbury had speculated that by refolding the proteins of viruses through chemical treatment we might well be able re-engineer them as textile fibres, and thus 'spin clothes from disease', as he memorably described it.[23] It was, he said, a suggestion that could have come from the dark imagination of Edgar Allen Poe and one that would probably have presented a formidable challenge to the marketing department of any clothing company that tried to bring such a fabric to market. There was, however, another raw material with great potential as an artificial textile fibre that presented no such problems.

At that time, monkeynuts (peanuts in the USA) were being grown in large quantities throughout the British Empire, with about 8 million tonnes being imported to the UK per annum.[24,25] By weight, they were composed about 50% of an oil that contained a globular protein called arachin. By extracting the arachin (the remainder of the extract was sold as cattle feed) and treating it with the compound urea, scientists at Imperial Chemical Industries (ICI) found that the delicate folding of the globular protein chain could be unravelled and refolded into an insoluble cream-coloured, crimped fibre.[26]

The chief attraction of this artificial fibre was its similarity to wool. A draft report by ICI described its properties as being that it could easily be dyed, it was resilient, soft, and warm to the touch, but, unlike wool, it was not attacked by moths.[27] It also showed elasticity and did not crease like cellulose fibres, nor did it show 'felting' like wool. Could this be an ideal new raw material from which to manufacture textiles? Shortage of wool for uniforms during the Second World War had shaken the textile industry out of complacency, and the possibility of producing textile fibres from peanuts promised a cheap, abundant raw material that, while it would not replace wool outright, might complement it and so enable British industry to fight off growing foreign competition.

The fibre also now had a name. ICI were so confident that the fibre would be economically viable that they established a pilot production plant and, as it was based at Ardeer in Scotland, its product was christened 'Ardil'. What was now needed was a practical demonstration of Ardil's potential as a textile fibre. The draft report on Ardil pointed out that it had a high capacity to absorb water and moisture and suggested that this property might make it the ideal textile fibre for underwear.[28] That this passage was later redacted in black marker pen suggests that, although delivering moisture-absorbent underwear to the nation might well hold great economic promise, it was presumably perceived as being too risqué for the more delicate sensibilities of the time. The British public were not yet ready to wear pants made from peanuts.

Far less problematic was an overcoat. As long-time consultant to ICI on fibre research, Astbury was sent an overcoat made from Ardil, the monkeynut protein fibre, produced at the Ardeer plant. But Astbury's connection with Ardil went much further than simply modelling its products, for it was to him that Ardil owed its very existence in the first place. The origins of Ardil lay in the work that he had done with A. C. Chibnall and Kenneth Bailey on the denaturation of globular proteins. One subject of this work had been denaturation by chemical treatment of a globular protein called edestin, which was obtained from hemp seeds. Astbury, Bailey, and Chibnall had shown that the denatured seed protein could then be refolded into a fibrous form, and they had filed patents on this process that were subsequently purchased by ICI.[29]

By the time that Astbury received his overcoat, ICI had not yet given any formal announcement of Ardil's existence, and they insisted that Astbury did not mention the fibre by name when giving lectures on the subject of artificial fibres.[30-33] It was a rather pointless restriction, since, as Astbury pointed out, 'every textile bloke in the country knows [about the Ardil fibre]', but, in the interests of maintaining his good relationship with ICI and protecting their commercial sensitivities, Astbury agreed to their requests to talk only in general terms until they were ready to make an official announcement about Ardil.[34]

When this announcement finally came in December 1944, Astbury was not best pleased. The source of his displeasure was the following passage in a copy of a draft of the announcement that was sent to him:

> starting metaphorically with a bag of monkey nuts and an idea, British chemists of ICI have evolved a new synthetic fibre ... The material is known as Ardil from the fact that the idea was developed and the experimental work carried forward at the ICI works at Ardeer—a factory founded by Alfred Nobel.[35]

Nowhere was there any mention of the original work done by himself, Chibnall, and Bailey. When Ardil was formally announced at a joint meeting of the Nottingham Section of the Society of Chemical Industry and the Royal Institute of Chemistry, on 14 December, Dr David Traill, an ICI research chemist, did acknowledge that 'Following the ideas of Astbury and Chibnall, our preliminary experiments were made ...'.[36] But Astbury felt that this barely gave credit to the role that he, Chibnall, and Bailey had played. Writing to John Weir, one of Traill's bosses at ICI, he declared:

> To be quite frank with you I must confess that I was surprised to find no mention in your official document of the part played by myself and other university people in the initiation of Ardil, if not in its actual production, and I cannot help feeling that the phrase 'Following the academic ideas of Astbury and Chibnall' in Dr. Traill's article in CHEM + IND might very easily be misinterpreted.[45]

The credit for Ardil was proving to be controversial. In response to several stories in the popular press regarding how coats from monkeynut proteins had been developed in Astbury's laboratory at Leeds and a lecture given by his colleague Professor J. B. Speakman that gave the same impression, ICI defended their position, saying:

> While in no way wishing to under-estimate Dr. Astbury's very valuable contribution to protein chemistry, we would protest that this statement [*press stories that Ardil was invented in Astbury's laboratory*] is not true. 'Ardil' is a registered trade name and the fibres produced under that name bear no resemblance to the fibres made by Dr. Astbury. His work with Professor Chibnall was principally on the protein edestin and they never experimented with ground nut proteins. Our use of ground nut proteins arose from our economic surveys of vegetable proteins.[38]

Feeling that their contribution had been both vital and sorely overlooked, Astbury was determined to set the record straight. In a letter to the editors of the journal *Nature* in December 1944, he declared:

> the impression created [*by the ICI announcement*] is that both the idea and the development are to be credited to I.C.I chemists, but the truth of the matter is that the basic idea came from me, as a pure deduction from my X-ray studies of protein structure, preliminary fibres were produced by my assistant K. Bailey, working in Prof. Chibnall's laboratory and then we three approached I.C.I on the matter.[39]

As far as Astbury was concerned, Ardil owed its very existence to having originated in academic research that was addressing fundamental questions of biology:

> I regard the evolution of regenerated protein fibres as one of the most striking examples of the century of the industrial application of science of the most fundamental kind.[40]

It was a point that he underlined when the editors of *Nature* invited him to contribute an article explaining the role that he, Chibnall, and Bailey had played in the story of Ardil.[41] In the letter, he argued that basic, fundamental academic science was essential to industry, yet it was deeply regrettable that such basic research should forever be having to demonstrate its practical worth.

The publication of Astbury's article in *Nature* finally pacified the issue, and all parties seemed happy to accept Astbury's account that, while he, Chibnall, and Bailey had shown how globular seed proteins could be unfolded and reformed into fibres by chemical treatment, it was the ICI research chemists who had then applied this process to monkeynuts for the production of Ardil. Dr Traill, who had made the official announcement of Ardil, wrote to Astbury, saying:

> I am very disturbed to learn that your feelings have been hurt by recent references to 'Ardil' and that my mention of the part played by Chibnall and yourself might easily be misinterpreted. I should be very sorry to think that I have in any way under-estimated your contribution since, in my opinion, and I have stated it in my lectures, you have done more than any of our contemporaries to arouse interest in the chemistry of proteins.[42]

Suitably placated by Traill's apology, Astbury wrote back declaring that he now considered the matter amicably closed and that he was sorry this disagreement had ever arisen:

> I propose therefore to wipe the slate clean and resume our former very friendly relations, including (especially on your part the telling of stories) and (exclusively on your part) the art of good eating and drinking.[43]

In 1946, ICI invested £2.1 million in a new production plant at Dumfries with the intention of producing 1000 tonnes of Ardil per annum. Marketed with the slogan 'Happy families with Ardil', the 1950s saw a worldwide push to promote this versatile new material that could be blended with wool to produce sweaters, blankets, carpets and felts or cotton and rayon for sport coats, dresses, shirts and carpets. As the price of wool was so high, Ardil seemed like a very cheap alternative, making it a very attractive raw material for textile companies.[44,45] It appeared that molecular biology had brought a new dawn for the British textile industry. In a letter to Traill, Astbury remarked that now was surely a:

> terrific opportunity to put the material on the market, for as far as I can see, any prospect of the price of wool falling seems pretty remote and I should say that by the time they do get the price down again, if they ever do, they will have been overtaken irretrievably by the onrush of artificial fibres.[46]

Astbury's skill, however, was as a physicist and not an economist. Just as Ardil came to market, the price of wool suffered a spectacular fall.[47] Another problem was that the supply of monkeynuts was not quite as abundant and easy to come by as had been hoped. The UK government at that time had given major backing to a plan called the East

African Groundnut Scheme, which aimed to turn vast tracts of barren ground in Tanganyika (now Tanzania) into fertile, agricultural land that could be used to grow monkeynuts on an industrial scale.[48] It was a plan that was both grand in its vision and equally grand in the scale of its failure. From the very start, the scheme was plagued with myriad difficulties, including poor soil, poor infrastructure, unreliable equipment, and even attacks from wild bees. As a result, the abundant and ready supply of monkeynuts hoped for never materialised.[49]

But aside from the falling price of wool and complications with the supply of groundnuts, the biggest problem of all facing Ardil was that it was simply not as good a material for textiles as ICI had hoped. As far as the local press were concerned, Astbury was firmly established as the man in the monkeynut coat, and, by 1952, most of his immediate family were wearing at least one garment made from Ardil fibres.[50,51] His wife and daughter knitted various jumpers and cardigans from fibres of the material, while his son had a cable-stitch pullover that he wore frequently before it fell victim to the new electric washing machine that the family had just bought and was shrunk to half its original size.[52] Astbury himself had been sent some material a few years earlier that he had since had made into a blue suit by a tailor in Leeds. Despite the sleeves being too short, on account of the material proving difficult to cut, he was very impressed at first. But, within a few months, the suit began to show considerable wear and tear, which got steadily worse over the course of the year that Astbury wore it, leading him to inform ICI of his conclusion that: 'Altogether, I am afraid the wearing quality of this suit was definitely inferior—in fact it eventually became more or less threadbare in the seat.'[53]

ICI finally ceased production of Ardil in 1957. Yet while it may have ultimately failed as a textile fibre, the genesis and production of Ardil nevertheless served as a powerful example of Astbury's conviction that molecular biology was concerned not just with observation of, but also the deliberate alteration of, biological chain molecules. Long before DNA was being chopped up with restriction enzymes and the scientists at Genentech were making hybrid DNA molecules to synthesise recombinant insulin, Astbury was sporting his monkeynut overcoat as proof that molecular biology could involve the deliberate manipulation of biological macromolecules for practical benefit.

In an intriguing case of art mirroring life, echoes of the Ardil story found themselves onto the cinema screen of the time. In 1951, around the same time that Astbury was complaining about the inferior quality of his increasingly threadbare monkeynut garments, the Ealing Film Studio released the film *The Man in the White Suit*, directed by Alexander Mackendrick. Nominated for an Academy Award for the Best Screenplay, it starred Alec Guinness in the main role as Sidney Stratton, an idealistic young chemist working in the laboratories of a textile mill in Northern England who invents a new synthetic textile fibre that is impervious to dirt and never wears out. At one point in the film, Stratton is trying in vain to explain his work to the mill owner's daughter played by the silky-voiced Joan Greenwood, and at first his questions about whether she understands the problem of polymerising amino acid residues or what a long-chain molecule is are met only with bewilderment. Then, gradually she begins to show some understanding and, as he sees the light finally dawn in her eyes, he cries in triumph 'That's it! Atoms stuck together—in this case like a long chain!' The line could have been written for Astbury.

The film was a light-hearted satire on the British class system of the time, but it also explored two themes that related to Astbury.[54] The first of these was to ask what science is for: is it an idealistic exploration of the workings of the physical world, or is it about utility—producing useful new products and gadgets? In the film, Stratton is convinced that his new fibre will revolutionise the textile industry and deliver millions of people from the drudgery of having to do laundry. But his noble intentions soon fall foul of both the capitalist mill owners who fear that his invention will destroy their profits and the mill workers who fear that it will destroy their jobs. Economic necessity takes precedence over Stratton's idealism—a point that is made quite clear in the film's climax when Stratton, clad in his white suit (which rather like Astbury's 'Ardil' suit is now starting to fall apart), is being pursued by both the mill owners and the mill workers, all of whom are adamant that his invention must never see the light of day.

Following his experience with Ardil, it was a tension with which Astbury could readily identify. In the aftermath of the First World War, Astbury's mentor William Bragg had been one of the voices calling for closer links between basic science and British industry, and, thanks to Bragg's conviction that X-ray crystallography could bring new insights to the textile industries of Leeds, Astbury had built a very successful career. But Astbury nevertheless felt that the relationship between what might be called fundamental, or academic, science and applied science, or technology, was a very delicate one. Writing to the Vice-Chancellor of the University of Leeds in 1955, he said:

> I appreciate that our livelihood as a nation depends very much on better and better technology and it is true that I myself have been fortunate in being able to contribute something to it, but all the same, I have to confess that it was always the quest into the nature of things that counted most. We have to do our utmost for technology, that is obvious, but please do not let us become too technological.[55]

For Astbury, science was first and foremost a 'magnificent adventure into the unknown' and an adventure not just of the intellect, but also of the spirit.[56] It was an act of enquiry into what he once called 'the Ultimate'. In this respect, he saw no distinction between science and the arts—both were equally ennobling pursuits that elevated the human soul:

> I feel much the same regard to the more ecstatic moments in science as I do with regard to music. I see little difference between the thrill of scientific discovery and what one experiences when listening to the opening bars of the Ninth Symphony. Inspiration has a common source, and great music, a great theorem by Newton, some bright idea that the lab-boy may run to tell you, a great speech in Shakespeare—all these are only aspects of one and the same thing.[57]

Science, he felt, was less a profession and more a calling, in which the scientist was a kind of craftsman, and, like the product of a skilled and gifted artisan, what distinguished this kind of work was its individuality as opposed to the uniformity of mass production.[58] But Astbury feared that this vision of science was under threat, partly thanks to its own success. The phenomenal power of science to solve problems and make new and useful products could very easily nurture a tendency to view it no longer as an end in itself, worth pursuing simply for its own sake, but rather as a means to an end. And herein lay the danger. For if science was allowed to become the mere slave of industry rather than its ally, it would cease to be the magnificent adventure that Astbury cherished and instead become a soulless utilitarian grind in which the sole concern was what he damningly called an 'eternal proneness towards judging progress by the number of registered patents, to the complete forgetfulness of what it all rests on'.[59,60] There would, he feared, be precious little room, if any, for 'the pioneer whose devotion is, by definition, to the as yet unrealised things, and whose name is nowhere to be found in a list of patents or a balance-sheet of commercial profits'.[61] Astbury despaired that, by becoming so utilitarian, the 'body of science is sick—sick almost unto death. It has become estranged from the spirit, the pure essence which alone can make it whole and sweet again'.[62] Despite once writing to Bernal that it was a job requirement for a molecular biologist not to be a romantic, this is exactly what he himself was, and maybe this sensitivity was rooted in a recognition that of all the sciences it was molecular biology, the very discipline that he had helped to found, that was particularly prone to being relegated to a mere industrial appendage.

According to some commentators, this is exactly the fate that befell molecular biology in the decades that followed Astbury's death. The writer Horace Freeland Judson said that molecular biology had once enjoyed a 'golden age', which he defined as:

> an age of innocence. It thrives, for a while, in the competitive harsh ocean of the twentieth century, as an island of idealism, and of play, and of, at the same time, an austere devotion to intellectual enthusiasm and openness ... Competition surely there will be, but competition that's more unifying of the tight small community than divisive—competition for approval, respect, and lively immediate intellectual response of your colleagues and mentors.[63]

This was the world inhabited by the likes of Astbury, Bernal, Chargaff, and Avery. In contrast, Judson observed that, by the 1980s, the golden age of molecular biology had given way to what he called an 'age of brass'.[64] The kind of scientific life enjoyed by Astbury, Preston, and Bailey had now been:

> superseded by a different sort of laboratory life and ethos: laboratories that are large and rigidly hierarchical, and an ethos driven by careerism, in which doing science is in large part a way to secure more grants, to get promotions, to gain power. Perhaps to get rich.[65]

Judson's observations were echoed by Erwin Chargaff, Astbury's fellow architect of molecular biology, who felt that genuine passion in science had now become replaced by naked raw ambition. Perhaps part of the reason for this was the very success of molecular biology—for with powerful examples like the production of recombinant insulin, this new science seemed to promise a cornucopia of novel medicines and disease therapies. Chargaff's fear was that such high expectations would actually be detrimental to the practice of science. Rather than pursuing science for its own sake, researchers would be forced instead to engage in a 'vigorous promotion campaign' as each group sought to emphasise how their work might cure cancer in an effort to compete for research grants. Chargaff despaired that in such an environment as this: 'What started as an adventure of the highest has become the survival of the slickest or the quickest.'[66]

But Chargaff's concerns went much deeper than merely lamenting the fact that, in his view, molecular biology had been degraded to a cynical rat race for patents, research papers, and grants, and they related to the second theme raised by *The Man in the White Suit*. As Sidney Stratton flees his pursuers during the film's climax, he dashes round a corner and nearly bumps into Mrs Watson, his old landlady and a washerwoman by trade. As Stratton begs her for some clothing to hide the white glow of his pristine suit that is now slowly disintegrating, she scowls and, fearing for what the suit means for the future of her trade, she responds in a tone that is at the same time, angry, bewildered and despairing: 'Why can't you scientists just leave anything alone? What's to become of my washing when there's no more washing to be done?'

Mrs Watson's concern may have been for her future as a washerwoman in the light of Stratton's new fibre, but her simple questions convey very succinctly the anxieties of the time about the power that the application of science could bring and, more importantly, how it could be misused. It was an anxiety that was shared by Chargaff, who, in his memoir *Heraclitean Fire*, pinpointed almost the exact date that he first began to feel it:

> It was an early evening in August, 1945—was it the sixth?—my wife, my son, and I were spending the summer in Maine, in South Brooksville, and we had gone on an after-dinner walk where Penobscot Bay could be seen in all its sunset loveliness. We met a man who told us that he had heard something on the radio about a new kind of bomb which had been dropped in Japan.[67]

Like Robert Oppenheimer reciting the words 'I am become Death, the destroyer of worlds' from the Hindu scripture, the Bhagavad-Gita, as he watched the nuclear fireball climb into the sky over the desert sands of New Mexico, the bomb had a profound effect on Chargaff.[68] It left him with:

> nauseating terror at the direction in which the natural sciences were going. Never far from an apocalyptic vision of the world, I saw the end of

the essence of mankind; an end brought nearer, or even made possible, by the profession to which I belonged.[69]

For Chargaff, the bomb was a potent symbol of a fundamental change that had occurred in the nature and culture of science. 'The time had long gone', he lamented, 'when you could say that you had become a scientist because you wanted to know more about nature.'[70] Science, he felt, was no longer just about making observations as Galileo had done looking through his telescope at the moons of Jupiter, or Darwin making notes on the fauna and flora of the Galapagos Islands. It was about gaining the power to actively change and shape the world. For Chargaff, this new power was not to be welcomed, or at least it was to be greeted with a necessary degree of humility and caution—and not only in nuclear physics but also in molecular biology. For if life could now be understood in terms of giant chain molecules and, moreover, these macromolecules could be deliberately altered, then did this not also raise the possibility that human beings now had the power to begin altering themselves?

It was a possibility about which Chargaff felt distinctly uncomfortable. If molecular biology became merely an all-powerful genie used by industry to sate the whims of a mass consumer society, then the consequences he warned might be dire:

> Once you can alter the chromosomes at will, you will be able to tailor the Average Consumer, the predictable user of a given soap, the reliable imbiber of a certain poison gas. You will have given humanity a present compared with which the Hiroshima bomb was a friendly Easter egg. You will indeed have touched the ecology of death. I shudder to think in whose image this new man will be made.[71]

The danger, for Chargaff, was that if human beings came to see themselves as nothing more than a complex assembly of giant molecules to be endlessly tinkered with and improved, then we would be left with an impoverished, dehumanised view of ourselves. Other founding figures of molecular biology were, however, less fearful. Like Chargaff, Bernal was equally appalled at the development of nuclear weapons, saying that they 'implicitly changed the whole existence of man in this universe' and that 'these weapons are inhuman in themselves ... their use cannot be tolerated whatever the excuse ... Atomic bombs are evil things',[72,73] but, for him, the insights from molecular biology that life could be understood as a system of giant chain molecules fitted well with the philosophical materialism at the core of his Marxism. Karl Marx had himself once declared that the job of philosophers was no longer to merely understand the world, but to *change* it; and it seemed that the role of scientists had now undergone a similar redefinition. For Bernal, molecular biology was simply another step in the inevitable story of human progress and the inexorable advance of reason and science:

> The practical scientists of today are learning to manipulate life as a whole and in parts very much as their predecessors of a hundred years ago were manipulating chemical substances. Life has ceased to be a mystery and has become a utility.[74]

He also believed that there already existed one shining example of what the future of humanity might look like given this power. Having vanquished perceived anachronisms such as religion and replaced them with a secular faith in reason, science, and inevitable progress towards an egalitarian future, the Soviet Union was for him the embodiment of this ideal. For Bernal, the USSR was a natural extension of what molecular biology promised—that not only biological molecules, but entire societies could be remodelled and re-engineered along efficient, scientific lines. With scientific research being directed and controlled by a central government allegedly for the public good of its people, Bernal, as well as many other British scientists of the time, were convinced that the Soviet model represented the inevitable and hopeful future of human civilisation.[75]

Astbury did not share Bernal's optimism. Science might show us how to deliberately untangle the chains of protein molecules, but Astbury was more realistic about what it could not untangle—the messy and complex reality of human beings, their societies, and their history. And while Bernal passionately believed that the Soviet Union was the first successful example of a society ordered along scientific lines, Astbury took a very different view:

> A cold-blooded scientific civilisation would be an abomination and such exercises in this direction as we have seen and are in fact witnessing even now seem to me sufficiently strong evidence already that human happiness cries out for something more. It would be rather boring in the

first place, but worse than that, with present-day notions of what constitutes science it would serve only to increase the unbalance and in time topple us over into the abyss.[76]

Given that the shadow of the bomb loomed so large at this time, Astbury's warning about being toppled over into an 'abyss' was a timely one. But he may well have been trying to express a more metaphysical danger to our humanity, and one that had been given a powerful fictional portrayal only a few years earlier.

In Aldous Huxley's novel *Brave New World*, his characters inhabit a perfectly safe, ordered society where every aspect of life, from the weather to childbirth, is controlled and regulated along rational, scientific lines. Everyone has a roof over their heads; no one goes hungry; the biological family unit has been abolished on the grounds of being an inefficient, out-of-date concept; centrally controlled technology provides for every need, and everyone gets exactly what they want. Everyone lives in material comfort and there is no loneliness, grief, anger, or depression because all these conditions, now deemed to be pathological, can be remedied with simple pharmaceuticals. Yet the inhabitants of Huxley's world seem less than human, living lives devoid of love, joy, curiosity, and reflection. As a consequence, there is no art, no music, no poetry, no literature, and no sense of aesthetic or moral values, all of which are deemed to have no practical utility. Liberated from these apparent delusions, the inhabitants of the 'brave new world' see themselves only in terms of their economic productivity, or else a bundle of primal appetites to be gratified and titillated.

Astbury once warned that, without great care, we might all be blown away 'by experiments carried out in a badly lit room'. Again, the most immediate interpretation that springs to mind is that this was anxiety about the bomb, but it could equally refer to the grim vision articulated so powerfully by Huxley.[77] For if the insights from molecular biology resulted in the dismissal of art, poetry, music, morality, and aesthetic values as little more than tricks of molecular assembly, then the floodgates might well be opened to the kind of dystopia that Huxley portrayed. In a paper delivered three years before his death, Astbury foresaw the powerful consequences that Watson and Crick's double helical structure of DNA would have on our understanding of genetics and our ability to manipulate life:

Recent discoveries in bacterial genetics are more than fascinating; they are also perhaps rather disturbing ... We can dimly see approaching a day when not merely bacteria but also higher animals and even human beings may have their transmissible characters moulded synthetically—a most alarming thought in some ways.

More dimly still, but not nearly so as it appeared only a short while ago, a day can be envisaged when we may be able, in a process analogous with infection by a bacteriophage, or with any fertilization, to add a synthetic polynucleotide to a collection of synthetic substrates and thereby set going a little hive of 'synthetic life'.

There is something seriously wrong with dreams such as this, of course. Systems of proteins and nucleic acids and accessory molecules with nothing else can scarcely keep on thinking and asking questions about themselves and experimenting on themselves until they have at last succeeded in making themselves. Reason apart, the halt will surely be called by an ultimate indeterminacy principle which says that it is impossible to place the required components in their correct places all at the same time.[78]

While Astbury would have welcomed the undeniable medical benefits brought about by the production of recombinant insulin, he would have been far more cautious about some of the philosophical interpretations of molecular biology and the recombinant DNA technology that grew from it. Accompanying this technology came a philosophical view in which human beings were simply the sum total of their genes. This genetic determinism was based on the view that genes are neat, discrete atomistic units of DNA to which every aspect of human behaviour must ultimately reduce. It was a view that was odds with Astbury's concept of the gene and one that, angering Chargaff's, made him refuse to accept that the subtleties and complexities of the human condition could reduce to mere quirks of base-pairing:

I cannot stomach people who claim that they have understood and explained 'Hamlet' by telling me how often the word 'and' appears in the first act. Is there a separate nucleotide code for your fingerprints which are different from mine? Is it a pairing error in position 79 which has produced the visions of Blake? It is above all against this shabby mechanization of our scientific imagination, which kills all ability to notice the unforeseen, that I protest, against this mat finish over a chaos of unrecognised ignorance, this butcher-like brutality with things that cry for gentle attention.[79]

In a lecture to an assembly of schoolmasters in 1948, Astbury voiced similar fears. His words lacked Chargaff's melodrama or rhetoric, but they were certainly no less powerful:

> Practical advances lead more and more to a mechanized civilisation, and it is right that for our material comforts and conveniences we should make the utmost use of machines. But they bring in their train an evil—indeed it is already amongst us. I refer to the gradual disappearance of the joys of handicraft and the pride with which the old craftsman could say, 'This is all my own work; I made it all myself'. There is nothing like creative achievement, be it never so humble, for human happiness and the health of the mind, and anything which tends to suppress it without offering nobler compensations is wrong and dangerous. Let us face the fact that there is an awful, aching emptiness growing upon us, a hunger for something we know not what. It is not simply the aftermath of war—if we think that we shall be deceiving ourselves. It is something which concerns our very nature, a fundamental sickness of the soul. Thoughtful people see it most clearly in the inadequacy and failing appeal of orthodox and authoritarian religion. No longer can we be fobbed off with any magical box of tricks or anything which forgets mankind's powers of reasoning, but—here's the rub—Nature abhors a spiritual vacuum even more than she abhors a physical one, and the void is beginning to ache, and ache badly.[80]

The 'void' of which Astbury spoke had its origins in the view that human life could simply be reduced to complex chemistry. Since the start of the twentieth century, the ongoing debate among scientists, philosophers, and intellectuals was about whether life could be reduced to a purely material phenomenon.[81] In his thought-provoking paper, 'Life, DNA and the model', historian Robert Bud argues that it was the emergence of molecular biology in the 1930s, 1940s, and 1950s that invigorated this debate. In his famous essay *The Two Cultures*, C. P. Snow said that the philosophical implications of molecular biology would be massive: 'This branch of science is likely to affect the way in which men think of themselves more profoundly than any scientific advance since Darwin's—and probably more so than Darwin's.'[82] The philosophical implications of a solely molecular view of life certainly polarised opinions. Leading the vanguard of an explanation of life in purely material terms were many pioneers of molecular biology such as Bernal and Crick. According to Bud, Bernal often liked to cite a quotation ascribed to Friedrich Engels that 'life was the mode of existence of albumen',[83] and, in his book *Of Molecules and Men*, Crick predicted that one of the effects of the rise in molecular biology was that:

> The old, or literary culture, which was based originally on Christian values, is clearly dying, whereas the new culture, the scientific one based on scientific values, is still in the early stage of development, although it is growing with great rapidity.[84]

While this was a development that Crick welcomed, others did not. In Bud's account, the main challenge to the mechanistic view of life that seemed to be at the heart of molecular biology came not from other scientists but from writers of fiction such as J. R. R. Tolkien and C. S. Lewis, whose imaginary worlds expressed their shared view that life transcended the purely material. When he gave the Riddell lectures at Durham University in 1943, Lewis spoke on the theme of 'The Abolition of Man', and expressed his concerns about what a reductionist, mechanistic view of life might mean for humanity's self-image. Lewis feared that a view in which human activity is a mere collection of molecules and mechanisms to be altered endlessly in the hope of improvement had the potential to leave human beings lost, with a diminished view of themselves and their world that would ultimately prove corrosive and destructive. If thoughts and feelings, beliefs and values reduced to nothing more than molecular interaction, if all human endeavour is to be measured solely in terms of utility, then surely any sense of morality, beauty, or meaning must disappear. Although more famous for his *Chronicles of Narnia* series written for children, it was in his novel *That Hideous Strength* that Lewis gave a powerful and vivid fictional expression to these fears. Central to the story is a scientific organisation called the National Institute for Co-ordinated Experiments (N.I.C.E), whose aim, according to one of its members is ultimately:

> for something better than housing and vaccinations and faster trains and curing the people of cancer. It is for the conquest of death: or for the conquest of organic life which sheltered the babyhood of mind the New Man, the man who will not die, the artificial man, free from Nature. Nature is the ladder we have climbed up by, now we kick her away.[85]

The utopia envisaged by N.I.C.E. in Lewis's novel, where technology has endowed humanity with the power to transcend its limits and gain complete mastery over its environment, resonated with the materialist philosophy of many of the pioneers of molecular biology like Bernal. It would, however, be a vast over-simplification to see this as a debate between polar opposites with battle lines drawn up neatly between reductionist scientists and writers of fantasy fiction. In the middle, there were characters such as Astbury, who, having spent his entire career advocating that life could be understood in terms of molecules and how they could be manipulated, felt very uncomfortable with what this might imply.

It was a discomfort with which Sidney Stratton might well have sympathised. When, at the climax of *The Man in the White Suit*, Stratton is challenged by Mrs Watson as to why scientists cannot just 'leave things alone', he has no answer. As she turns her back on him and walks away, Stratton can only stand in silence and watch her heading down the street into the darkness. As he does so, his eyes have a faint look of guilt in them, knowing that his science may well have cost the old lady her livelihood.

Like Stratton, Astbury received a similar challenge. In response to one of his BBC radio broadcasts in which he cited the monkeynut coat as an example of the power of science to transform the world for the better, a listener wrote to him pointing out that material progress by no means guaranteed moral or ethical advancement, and certainly not utopia.[86] Unlike Stratton, however, Astbury did not remain silent, but wrote the following reply:

> It is true that part of my laboratory is devoted the study of industrial fibres, but the other part, the part nearest my heart is devoted to the study of the molecular structure of biological tissues . . . The great lesson of my talk, as you perceived, was that all proteins, including the viruses are similarly constituted, and all living things are fabrics of thread molecules, so that there is nothing at all fantastic in making an overcoat out of monkey-nuts, or even out of a disease. The physical components of living matter are always the same—we eat steak to grow whiskers for example—BUT it does not follow that I personally conclude that there is nothing else to life but intricate arrangements of molecules. My experiments and reason tell me that it certainly looks as if all behaviour can be explained in terms of the physics and chemistry of molecules, but inside me I reject that view. I feel there is something much greater behind it all—don't ask me what. I don't know.
>
>   I just hope—and like you, I want to be just happy. I am afraid I have never succeeded in being really happy for a long time—no one ever does

succeed, I suppose, but I agree with you, as every right-minded person must, that ordinary simple happiness is the thing to look for. Probably, and unfortunately, it is beyond my reach because I know too much but I know well enough what you mean. <u>Children</u> know what you mean, and the important thing is not to grow up.

Please don't think too severely of science, I beg of you. True science is a wonderful thing, and my work is the most wonderful thing about me—the most wonderful thing that I know. It is endlessly gratifying, endlessly moving to search day after day, night after night into the nature of things—into the nature of our very selves. The science I know and practise is the crowning glory of human culture. I classify it with music and literature, and every other beautiful thing in life. Believe me, science is not just bombs, luxuries, wireless and what-not. It is man's greatest intellectual adventure, his journey into the unknown.[87]

From its unlikely beginnings with research into wool, Astbury's own journey into the unknown had led him to DNA. But although he and Florence Bell had shown that X-rays could be used to reveal the regular, ordered structure of DNA and in so doing paved the way for Wilkins, Franklin, Watson, and Crick, they had also done much more. For as Nobel Laureate Max Perutz pointed out, thanks to the field of structural biology that had emerged from Astbury's work we could now understand:

why blood is red and grass is green ... how muscles contract, how sunlight makes plants grow and how living organisms have been able to evolve into ever more complex forms ... The answers to all these problems have come from structural analysis.[88]

Had Perutz been writing those words today, he might well have added one more example to his list. It is one that has, sadly, become a household name of late. Thanks to the now infamous spike glycoprotein that protrudes from its surface, the SARS-Cov2 virus is able to bind to and infect human cells.[89] It is able to do this because the precise three-dimensional configuration into which the amino acid chain of the viral spike protein is folded allows it to physically dock with certain specific proteins on the surface of cells in the human upper respiratory tract. When this happens, the spike protein undergoes a change in its conformation that then allows the virus particle to fuse with the membrane of the human cell and deliver the viral genetic material into the new host where it can replicate to form copies of itself.[90]

Astbury's vision of structural biology had begun with an explanation of why wool could be stretched and a coat could be woven from monkeynuts, but today it allows us to understand how specific mutations in the Sars-CoV-2 spike protein can alter its shape to make the virus more infectious.[91] This knowledge in turn helps in the design of vaccines to block the action of the virus. Astbury's monkeynut coat, spun from an ingenious act of molecular origami, may have long since rotted away in a dusty cupboard, but the idea of structural biology from which it was woven has left a lasting and powerful legacy. And that alone surely makes its story, along with that of the man who wore it, worth telling.

Florence Bell (1913–2000)—The 'Housewife' with X-ray Vision.

*Based on a paper first written for Notes and Records of the Royal Society (2021),
'On the Shoulders of Giants' a letter published in the journal 'Inference' in August
2021, and the Girton College Annual Alumni lecture, 25 September 2021.*

In Simon Armitage's wonderful poem, 'Ten Pence Story', the UK poet laureate gives voice to an ordinary ten pence coin as it reflects sadly that it will never realise its dream of deciding the toss at a Wembley cup final but will instead be melted down into oblivion and anonymity. Had Armitage chosen the new fifty pence coin released in July 2020 by the Royal Mint as his subject, however, his numismatic narrator might have had a far less bleak outlook about their fate.

Released to commemorate what would have been the one hundredth birthday of the scientist Rosalind Franklin, the coin is inscribed with 'Photo 51' and is just one of a number of accolades that she has accrued in recent years. Others include a 2015 hit West End play for which the actress Nicole Kidman earned an award for her portrayal of Franklin, and most recently a rover that will touch down on Mars in 2023 and drill into the surface of the Red Planet in search of life.

We can only speculate about what Franklin would have made of such stardom, for she herself seems to have had a rather modest appraisal of her role in the discovery of the DNA structure. When first told of Watson and Crick's success, she is said to have simply remarked, 'We all stand on each other's shoulders'.[1] This was most likely a reference to Sir Isaac Newton's famous line about having seen further only by standing on the shoulders of giants. But rather than being original, Newton himself was most likely citing a reference by the 12th-century French philosopher Bernard of Chartres to Orion, the giant hunter from classical mythology who was blinded as a punishment and carried the dwarf Cedalion on his shoulders as his guide.

Regardless of its origins, however, it was certainly an apt remark for Franklin to make, for she too had stood on the shoulders of someone else. With her demonstration that the regular, ordered structure of DNA could be revealed by X-ray crystallography, Florence Bell had paved the way for Franklin and Raymond Gosling to take 'Photo 51'. And had she been able to remain in Astbury's laboratory, she might have done much more.

For although Astbury was a giant in his field, like the mythological hunter Orion, he too suffered a kind of blindness. Astbury's friend and colleague Reginald Preston once reflected that Astbury 'was always fired by the prospect of solving a biological problem through molecular architecture' but added that 'his first flush of enthusiasm soon faded if the answer was not forthcoming in a field which did not bear directly on his own views of protein structure'.[2] For Astbury, molecular shape was everything. 'Configuration is all', he once wrote, 'A polypeptide chain in one configuration is a different molecule from the same chain in another configuration'.[3]

This may go a long way to explaining why Astbury never grasped the importance of Beighton's photo for solving the structure of DNA. In a paper describing a short-lived foray by Astbury into the field of embryology, Professor Jan Witkowski of Cold Spring Harbor Laboratory, points out that Astbury's insistence on the primacy of three-dimensional structure limited his vision severely. And while the blind Orion at least had the dwarf Cedalion sitting on his shoulders to act as his eyes, Astbury had no such guide.

This might not have been the case, however, had Astbury's pleas to the War Office for Bell to be exempted from War Service not fallen on deaf ears. For had Bell been able to remain in Astbury's lab, it may well have been her—and not Elwyn Beighton—who ended up taking the stunning new X-ray images of DNA that were so similar to 'Photo 51'. And it is highly unlikely that an X-ray crystallographer of Bell's calibre would have been content to let Astbury simply file away photographs of such striking quality in a drawer to be forgotten. And with Bell's formidable intellect and her willingness to challenge the limitations of Astbury's vision, his curiosity might have been piqued to investigate those photographs further. And even had Bell herself not been able to solve the structure of DNA from these photographs, she might well have spurred Astbury into showing them to Linus Pauling when he visited Astbury at his home in Leeds in 1952.

In his excellent book *Unravelling the Double Helix*, Gareth Williams says that 'A year later and in another place, Beighton's photographs would have created a sensation ... Astbury did not know what to make of them'.[4] But what if, thanks to Bell's encouragement, Astbury had known 'what to make of' these images? Might he really have gone on to be remembered as the discoverer of the double-helical structure of DNA? This is one possibility considered by Matthew Cobb, historian and Professor of Zoology at the University of Manchester, who has proposed a number of plausible alternative historical routes by which the structure of DNA might have been discovered.[5]

But even had Astbury grasped that DNA had a double-helical shape, this discovery would have offered him no clues to the function of the molecule, in the same way as it did for Watson and Crick, who had grasped the crucial insight that a double-helical structure allowed the pairing of complementary bases on opposite strands and that this pairing allowed the molecule to replicate itself and copy genetic information. Moreover, in a second *Nature* paper published six weeks later, Watson and Crick proposed what Cobb has called their 'brilliant suggestion' about how the molecule might actually carry this information in the first place:[6]

> it therefore seems likely that the precise sequence of the bases is the code which carries the genetical information.[7]

This insight owed much to the work of Erwin Chargaff and his application of Archer Martin and Richard Synge's method of partition chromatography to the analysis of DNA which had shown that the composition of bases varies between species. In *The Eighth Day of Creation*, his majestic and exhaustive history of molecular biology, Horace Freeland Judson has hailed this achievement as being 'the discovery that made molecular biology possible and for which we must honor Erwin Chargaff'.[8] But it is somewhat ironic that this discovery was made by Chargaff, and not Astbury, for not only were Martin and Synge working literally down the road from Astbury's laboratory but they also collaborated with him in his own work on wool fibres. On one occasion Astbury even wrote with delight that 'we are rushing together in a mutually helpful manner' and declared with excitement 'that we are on the verge of something epoch-making in protein studies'.[9] But his correspondence from this time also betrays a sense that he saw Martin and Synge's work as being very much subservient to his own X-ray studies of the three-dimensional structure of wool proteins. For Astbury, molecular biology was always rooted in understanding the three-dimensional structure of proteins and DNA.[10]

Had Bell remained on the scene, events might have unfolded differently. Bell had co-written a paper with Martin and Synge, and later in her life recalled with pride about how she had worked with two scientists who had gone on to win the Nobel Prize.[11] Had she remained in Astbury's lab, she might well have encouraged him to consider how partition chromatography could be applied to DNA.

But by this time, DNA was all in Bell's past. After working as an industrial chemist for the Magnolia Petroleum Company in Beaumont, Texas, she gave up her career to look after her four adopted children. It was presumably in reflection of these changed circumstances that, when she died in 2000 her occupation was simply described as having been a 'housewife'—with no mention of her previous scientific career.[12]

Bell herself always joked that one of her greatest achievements was possibly to have been the first woman in the RAF to have worn trousers.[13] She was being very modest. When William Bragg had first left Leeds for London in 1915, he had urged that the Textiles Department find itself 'a keen young man' who could apply his Nobel Prize-winning discovery of X-ray crystallography to the study of wool. But what he could never have anticipated was that it was a keen young woman who would take this work beyond wool to the study of DNA, for with her demonstration that the regular, ordered structure of DNA could be revealed using X-ray crystallography, Bell had paved the way for Rosalind Franklin's commemorative coin and Mars rover.

Acknowledgements

With special thanks to Bell's son, Mr. Chris Sawyer, for all his help and support with this work.

Laszlo Lorand (1923 – 2018)—The Biochemist Who Came in from the Cold

(A version of this article first appeared in *The University of Leeds Alumni* magazine, 2021)

Introduction

When Laszlo Lorand, a young scientist working at the Institute of Biochemistry in Budapest, was summoned to the office of his supervisor Bruno Straub on the afternoon of Friday 17 December 1948, he presumed it was for just another routine discussion about the progress of his research.

Lorand enjoyed his work at the Institute, where life at the lab bench was enlivened with fiercely competitive games of volleyball and table tennis, and its director, Nobel Laureate Albert Szent-Györgyi, offered his staff free tickets for concerts and operas. But earlier in 1948, he had received an invitation from Astbury to come and work in his lab in Leeds. For a young scientist like Lorand who was just starting out in his research career, this was too good an opportunity to turn down. But as he was about to discover, the authorities in his native Hungary had very different feelings.

Two years earlier, in the same year that Lorand had joined the Institute as a third-year medical student, Winston Churchill had made his grim observation in a speech given at Fulton, Missouri that 'an iron curtain has descended across the continent.' Now Churchill's 'iron curtain' was about to come crashing down on Lorand with full force. As the Cold War began and former wartime allies now became adversaries, the authorities in Eastern Europe did not look favourably on their young scientists travelling to the West. Having begun the meeting by praising Lorand's work, Straub informed him that his passport would be withdrawn by the Communist authorities on the following Monday morning. There was to be no trip to Leeds.

As a survivor of the Holocaust, Lorand refused to become the victim of yet another brutal totalitarian regime. That weekend he made a decision that would change his life and, most likely, break his heart. On Sunday afternoon he said goodbye to his widowed mother and fiancée and, with only £5 in his pocket, boarded a train bound for Vienna. Once he was safely out of the Communist Bloc, he made his way through the ruins of post-war Europe, paying his way with cigarettes, eventually arriving in London.[1]

On 2 January 1949, Lorand finally arrived in Leeds. His single suitcase had been packed by his mother, who, fearing that Britain was still suffering food shortages, had stuffed it full of a freshly roasted goose and a large stick of hard salami.[2] It was in Leeds that Lorand resumed his research into a subject that had fascinated him since his days as a medical student: how blood clots form.

While studying medicine, Lorand had been amazed at the near miraculous sight of blood plasma being turned (within a matter of only minutes) from liquid into a solid clot merely by the addition of a few drops of liquid containing the blood protein thrombin.[3] This rapid transformation had important medical implications, for while clot formation was essential for healing at sites of injury, if it occurred elsewhere in the body, it could result in a fatal thrombosis, or circulatory blockage. Yet, despite its physiological importance, the mechanism of how thrombin caused clot formation remained a mystery—and one that Lorand had hoped to solve when he had first joined Budapest's Institute of Biochemistry.

Working in the spartan conditions of Astbury's new department—with its unreliable electricity supply, risk of flooding, and wobbling scientific instruments—Lorand began his studies. He made regular visits to either a local abattoir or to the Red Cross, from which he would return, sitting on the tram with a bottle of blood clutched tightly to his chest to avoid spilling the precious material. But adverse working conditions were the least of his concerns. Within a year of his arrival in Leeds, he received a letter from Straub instructing him to return to Hungary at once. To the Communist authorities, the idea that such a promising young scientist as Lorand could choose of his own volition to remain in the West was an unacceptable loss of face—and one which they would not allow.

Astbury was quick to come to Lorand's defence and argued to both the University and Hungarian authorities that his work was so important that he should be allowed to remain:[4]

As far as I can see, the only 'crime' that Lorand has committed is that he was rash enough to leave Hungary to work in this University and has therefore fallen under the bane of Western influences. In these days, as you know, that can be sufficient to damn anyone from the Eastern European countries, and it appears that it has indeed been sufficient to damn Lorand.[5]

Although now stripped of his Hungarian citizenship, Lorand was able to remain in Leeds. Thanks to Astbury's intervention, he went on to make the ground-breaking discovery that thrombin acts as an enzyme to convert the soluble blood protein fibrinogen into insoluble fibrin—the active agent of clot formation.[6,7]

Lorand's discovery earned him a PhD and was also quickly recognised as a landmark in understanding the mechanism of blood clot formation. After leaving Leeds to move to London's Lister Institute of Preventive Medicine, Lorand then made a second major discovery when he isolated Factor XIII, a blood component vital for clot stability and which is involved in certain rare hereditary blood disorders.

Shortly afterwards, Lorand moved to the USA, but he stayed in touch with Astbury and, as Soviet tanks rolled through the streets of Budapest during the 1956 invasion of Hungary, Astbury wrote to express his support for Lorand and the family he had left behind there:

> We have often thought of you in relation to this horrible business in Budapest, [. . .] we were very sorry to hear that your own family were unable to escape.[8]

Music had provided consolation in dark times for both Astbury and Lorand, and while in Leeds both had often enjoyed playing duets on the violin together. In a letter of 1957, however, Lorand admitted to Astbury that his latest attempts to learn the piano had been far from successful:

> I played Auld Lang Syne on New Year's Eve at midnight. Nobody applauded, in fact, I seemed to dampen the merriment.[9]

In 1963, Lorand took his young family on a visit to Europe and, for the first time since that fateful Sunday afternoon in 1948, when he had stepped onto the train from Budapest, he was finally reunited with his mother, who had been granted special permission to leave Hungary in order to make the trip.[10] Fearing that it might offer a chance for them to flee to the West, however, the Hungarian authorities had denied other family members from making the trip with her. It would be another thirty-three years before Lorand finally returned to visit his homeland of Hungary, by which time the Iron Curtain had lifted and the Cold War was consigned to history.[11] By this time, he was established as a distinguished figure in the field of research into blot clotting—research that not only revealed the mechanism by which blood clots form, but also in developing anti-clotting treatments. For one young medical student, Monty Losowsky (1931–2020), this was of particular importance. Losowsky was very the same young doctor who, having tried—and failed, initially at least—to inject Astbury with an anti-clotting agent, had been dispatched to the medical library with strict orders not to return until he could give an account of how blood clots form. On his return, Losowsky had taken Astbury's lesson to heart—he went on to establish the first department of molecular biology at the University of Leeds School of Medicine. After the disappointments that Astbury had faced when trying to establish his own national centre for molecular biology at Leeds, he would have been delighted to see that the vision he had proposed in the aftermath of the Second World War was finally becoming a reality. And after his lonely trek across war-ravaged Europe with only a suitcase at his side, so too, no doubt, would Lorand.

Acknowledgements

I express my thanks to Professor Laszlo Lorand (1923–2018) for sharing with me a draft of his unpublished memoirs as well as a very enjoyable correspondence during the last few years of his life. I also thank his daughter Dr. Michele Lorand for her welcome support and for kindly sharing her memories of her father. I am also very grateful to Professor Monty Losowsky (1931–2020) for his discussion of Lorand, Astbury, and their work together.

APPENDIX C

George Washington Carver – 'The Peanut Man'?

Imperial Chemical Industries (ICI) may well have made Astbury a coat woven from monkeynut fibres, but they were certainly not the first to recognise that commercial gold might lie within the husk of the humble peanut, and attempt to exploit this possibility. Writing in 1917, the US agricultural scientist George W. Carver (1864-1943) had sung the praises of peanuts, describing them as having 'almost limitless possibilities'.[1] Born as a slave in Missouri around 1864, Carver had risen to become Director of the Tuskegee Institute in Alabama where his research was directed towards developing practical help for impoverished small family farms in the rural Southern US, many of which were run by former slaves or their descendants. Carver recognised that dependence on growing only cotton, as had been the practice for decades in the Southern American states was detrimental for two reasons. Firstly, it kept the farmers impoverished as they had to buy expensive commercial fertilisers and cotton seed from large companies and secondly, the continuous growth of cotton on a large scale left the soil seriously depleted of nutrients. To address this problem, Carver championed sustainable farming methods such as crop rotation and the use of organic fertilisation methods. Carver also encouraged these farmers to break with tradition and diversify their crops to include not just cotton but also plants which would replenish the soil. Of these, one of Carver's favourites was the peanut. These offered not only a cheap and abundant source of dietary protein for the rural poor, but also a means of lifting them out of poverty whilst at the same time improving the soil. In one of his famed bulletins written for local farmers, Carver listed 105 recipes made using peanuts – but he soon realised that their potential might go much further.[2] When, thanks to his expertise, he was called upon in 1921 to testify to the US House of Representatives Ways and Means Committee on import tariffs for foreign peanuts, Carver argued for the superiority of the home-grown US peanut and was already thinking about how it might have commercial potential beyond simply the food industry.[3] Four years later, he filed a patent on cosmetic creams derived from peanut oil and, according to the website of Tuskegee University, had proposed a list of commercial products including paints, shampoo and even nitroglycerine – all made from peanut extracts.[4,5]

As a result, Carver soon achieved national fame as 'The Peanut Man', but in her biography, *George Washington Carver: Scientist and Symbol*, the historian Linda McMurry argues that this moniker is largely a mythological construction.

Moreover, it is one that has eclipsed Carver's real legacy.[6] Due to the rise of industrial agriculture in his own time, Carver's evangelism for sustainable farming that reflected and harmonised with the local natural environment had met with little traction, but with growing concern today about environmental damage, his ideas have taken on a new urgency and relevance.[7,8] And perhaps it is for this, more than the possibility of shampoo from peanuts, that he should be remembered.

Notes and References

Abbreviations

ULSC BL   University of Leeds, Special Collections, Brotherton Library

Introduction

1. (Wilkins 2003); p.137.
2. (Olby 2009); p.129.
3. Watson, J. D. (1968) *The Double Helix*. London: Weidenfeld & Nicholson, p.21.

Chapter 1: A Picture Speaks a Thousand Words

1. Watson and Crick, 1953b, pp. 737–738.
2. Watson, 1968: p. 21.
3. Wilkins, 2003, pp. 218–219.
4. Watson, 1968; p. 167.
5. Watson and Crick, 1953a; pp. 964–967.
6. Devlin, H. (2013) 'My sister, her DNA breakthrough and Nobel prize that never was', *The Times*, 5 June.
7. Watson and Crick, 1953b; p. 738.
8. Letter from Francis Crick to Jacques Monod, 31 December 1961. London, Wellcome Library, PP/CRI/H/3/5/1. Online at: 'Profiles in Science', The Francis Crick Papers, National Library of Medicine, <http://profiles.nlm.nih.gov/ps/access/SCBBFW.pdf>. Cited in Glynn, 2012; p. 156. Cited with the kind permission of the Wellcome Library.
9. Maddox, 2002; p. 316.
10. Maddox, 2002; p. 325.
11. Letter from F. H. C. Crick to J. D. Watson, 13 April 1967. London, Wellcome Library, PP/CRI/I/3/8/4. Cited in Glynn, 2012; p. 155. Cited with the kind permission of the Wellcome Library.
12. Watson, 1968; p. 18.
13. Watson, 1968; p. 166.
14. Watson, 1968; p. 17.
15. Watson, 1968; p. 69.
16. Maddox, 2002; p. 312.
17. Sayre, 1975.

18. Sayre's view that the problems encountered by Franklin in working with Randall and Wilkins were due to a culture of innate male chauvinism at King's College has, however, since been challenged. In *The Eighth Day of Creation*, his exploration of the history of molecular biology, Horace Freeland Judson recounts a number of interviews with female scientists working at King's during this period, and notes that none of them recalled feeling a sense of oppression on account of their gender. Moreover, Judson's interviews reveal that, far from being a bastion of male chauvinism, King's actually had a higher proportion of senior female scientists than most other institutions at that time. On the basis of these interviews, Judson concludes that 'those of Franklin's colleagues at King's who were women unanimously reject the view that her troubles there arose because she was shut out as a woman. Responding decades after the events, these women know that professional women often do face barriers, systematic barriers that men do not meet. None the less, they reject as unhistoric and anachronistic the use of Rosalind Franklin as an emblem for the condition of women in science.' From 'In defense of Rosalind Franklin: The myth of the wronged heroine' in Judson, 1996; pp. 625–627.

19. Maddox, 2002; p. xviii.

20. Judson, 1996; p. 629.

21. Glynn, 2011; pp. 157–160.

22. 'I know more about cancer than cancer specialists', James Watson interviewed by Giles Whittell, *The Times*, 10 June 2013.

23. Klug, 1974.

24. Maddox, 2002; p. 202.

25. Ibid.

26. Ibid.

27. Maddox, 2002; p. xviii.

28. Glynn, J. (2013) 'DNA's heroine', Letter to the Editor of *The Times*, 11 June.

29. Fara, 2009; p. 317.

30. Mention should also be made here of Freda Collier, who was Franklin's X-ray photographer and headed the photographic laboratory at King's College that produced 'Photo 51'. According to an obituary by her niece, Collier was a close confidante of Franklin, and it was a great regret to her that Franklin's contribution to Watson and Crick's discovery was never fully recognised at the time. ('Freda Collier', *The Guardian*, 9 February 2013.)

31. Glynn, J. (2011) 'My sister Rosalind Franklin', Brodetsky Lecture delivered at the University of Leeds, 6 November 2011. Personal communication. When Newton made his now often quoted comment to Robert Hooke about having stood on the shoulders of giants, he was not himself being original but was most likely paraphrasing a reference by the 12th-century French philosopher Bernard of Chartres to the legend of Orion, the giant-sized hunter in classical mythology who, having been blinded as a punishment, carried the dwarf Cedalion on his shoulders as his guide. See Merton,

R. K. (1965) *On the shoulders of giants: a Shandean postscript.* Chicago: University of Chicago Press. When first told of Watson and Crick's success in solving the structure of DNA, Franklin is said to have remarked that 'We all stand on each other's shoulders'. See Williams, G. (2020) *Unravelling the double-helix,* Weidenfeld & Nicholson, p. ix. See also 'The Secret Lives of DNA' by Neeraja Sankaran, *Inference,* 6(1), June 2021 for an interesting and alternative interpretation of Franklin's comment. https://inference-review.com/article/the-secret-lives-of-dna

32. Maddox, 2002; p. 121.
33. Letter from Warren Weaver to W. T. Astbury, 27 May 1948. Astbury Papers, MS419 Box E.153, ULSC BL.
34. Bernal, 1963; p.1.
35. Ibid; p. 29.
36. Bailey, 1961; p. xiv.
37. Preston, 1974; p. 1.
38. Preston (1974), p. 19.

Chapter 2: 'Germany Has Much to Teach Us . . .'

1. Henderson, M. (2009) 'Ian Botham's Headingley 1981 Ashes Test still resonates', The Daily Telegraph, 9 August; <http://www.telegraph.co.uk/sport/cricket/international/theashes/5984229/Michael-Henderson-Ian-Bothams-Headingley-1981-Ashes-Test-still-resonates.html.>
2. For an excellent history of Headingley, see Bradford, 2008.
3. Astbury, 1942.
4. Bennett, 1908; p. 39.
5. Bailey, 1961; p. xi.
6. Bennett, 1908; p. 39.
7. Bailey,1961; p. xi.
8. Ibid.
9. Bennett, 1902.
10. Chibnall, 1964.
11. Army Form B. 2512A. Certificate of Attestation for William Thomas Astbury, 30 May 1916; Army Form B. 2079 Certificate of Discharge. Regimental No. 116104; Rank Lance Corporal; Regiment 16th Royal Army Medical Corps; 18 February 1919. These discharge papers also mention that on 17 February 1919, Astbury underwent an operation at the Central Military Hospital in Cork for a ventral hernia, which was a condition arising from an operation for appendicitis in June 1915. Private papers of Mr W. Astbury. Cited with kind permission of Mr W. Astbury.
12. Bill Astbury, grandson, personal communication.
13. Röntgen 1896. Translated into English by Arthur Stanton from *Sitzungsberichte der Würzburger Physikalisch-Medicinische Gesellschaft,* 1895.

14. Some historians have challenged the idea that X-rays were universally and quickly adopted by the medical profession. See for example, Jamieson, 2013.

15. Jones and Lodge, 1896.

16. Rossi and Kellerer, 1995.

17. Preston, 1974; p. 2.

18. Olby, 1994; p. 46.

19. Jenkin, 2011; p. 70; see also S. Talbot Smith, 'Memories of Sir Wm. Bragg', *The Mail* (Adelaide), 4 April 1942, p. 7, cited in Jenkin, 2011; p. 71.

20. Caroe, 1978; p. 29.

21. Letters W. H. Bragg to Council, 14 May, 16 November, and 15 December 1886, UAA, S200, dockets 171/1886, 447/1886 and 506/1886, respectively. Cited in Jenkin, 2011; p. 84.

22. 'Report on a lecture given by Sir William Bragg at the Royal Institution, London on 'The early history of X-rays', *Nature*, 123(1929), p. 218.

23. 'Rontgen photography unsuspected', *South Australian Register*, 30 May 1896, p. 5; <http://nla.gov.au/nla.news-article53707190>.

24. Ibid. See also Caroe, 1978; p. 38.

25. Rontgen rays. *South Australian Register*, 5 June 1896, p. 5; <http://nla.gov.au/nla.news-article53688180>.

26. 'The Rontgen photography', *South Australian Register*, 18 June 1896, p. 6; <http://trove.nla.gov.au/ndp/del/article/53695446>.

27. W. L. Bragg, autobiographical notes, pp. 6–7, cited in Jenkin, 2011; p. 147.

28. Jenkin, 2011; p. 218.

29. Thornton, 2002; p. 171.

30. Letter from L. Scammell dated 30 September 1931 to Dr Arthur Lendon. Available at website of the South Australian Medical Heritage Society <http://www.samhs.org.au/VirtualMuseum/xrays/FauldingsandXRays/FauldingsandX-Rays.html>. See also 'Rontgen photography unsuspected' (n 23).

31. Gilleghan, 2001.

32. 'X-rays outside the laboratory', lecture given by Dr A. Jamieson, 23 March 2013, Bragg Centenary Day, University of Leeds.

33. Caroe, 1978; p. 52.

34. Bede, *Ecclesiastical History of the English People* (translated by Leo Sherley-Price, first published 1955; revised edn (1990) R. E. Latham; Introduction and Notes by D. H. Farmer). London: Penguin Books, pp. 132 and 184.

35. Thornton, 2002; pp. 7–10.

36. Burt and Grady, 1994; p. 12.

37. Herman von Puckler, Muskau Prince, *Tour in Germany, Holland and England, 1826, 1827, 1828* (1832), IV, p. 210. Cited in Burt and Grady (1994), p. 87.

38. From Toft, J. (1966) *Public health in Leeds in the nineteenth century: a study in the growth of local government responsibility c. 1815–80.* Unpublished MA thesis, University of Manchester, p. 103. Cited in Morgan, C. J. (1980) 'Demographic change, 1771–1911', in Fraser, D. (ed.) *A history of modern Leeds.* Manchester: Manchester University Press, p. 46.

39. Thornton, 2002; pp. 145–146.

40. *The century's progress: Yorkshire industry and commerce.* London Printing and Engraving Co. (1893), p. 150. Cited in Burt and Grady, 1994; p. 133.

41. Nelson, 1980.

42. Thornton, 2002; p. 32.

43. Heaton, 1965; pp. 1–2, and Chapter 1. Cited in Burt and Grady, 1994; p. 19.

44. Forster 1980), p. 8.

45. Nelson (1980), pp. 4–5.

46. Heaton (1965).

47. Burt and Grady (1994), p. 36.

48. Burt and Grady (1994), p. 35.

49. Defoe, D. (1724–26/1971) *A tour through the whole of Great Britain*, P. Rogers (ed.). London: Penguin, pp. 500–505. Cited in Burt and Grady, 1994; p. 53. See also Nelson, 1980; p. 21.

50. Burt and Grady, 1994; p. 63.

51. Ibid; p. 57.

52. Ibid.

53. Nelson, 1980; pp. 42–43.

54. Caroe, 1978; p. 65.

55. Ibid; p. 53.

56. Ibid.

57. Ibid.

58. Caroe, 1978; p. 65.

59. Ibid; p. 66.

60. Bragg, 1912.

61. Letter from W. H. Bragg to E. Rutherford, 5 December 1912, CUL RC B392. Cited in Jenkin, 2011; p. 334.

62. Phillips, 1979; p. 90.

63. Bragg and Bragg, 1913.

64. Jenkin, 2011; p. 339.

65. Caroe, 1978; p. 75.

66. Phillips, 1979; p. 91.

67. Caroe, 1978; pp. 75–8.

68. Jenkin, 201; p. 366.

69. Caroe, G. (1976) 'Bragg, Gwendoline', in Biven, R. (ed.) *Some forgotten . . . some remembered: women artists of South Australia.* Norwood: Sydenham Gallery. Cited in Jenkin, 2011; p. 366.

70. Letter from W. H. Bragg to E. Rutherford, 12 May 1920, RI MS WHB 11A/24; see also letter W. H. Bragg to President, Swedish Academy, 17 May 1920, RI MS WHB 11A/25. Cited in Jenkin, 2011; p. 401.
71. Simmons, 1910; p. 313.
72. Pohl, 1909; p. 205.
73. 'Manchester's Report on Technical Education in Germany and Austria', (1897) *Nature*, 56, pp. 627–630, 628.
74. Ibid, p. 630.
75. Pohl, 1909; p. 206.
76. Ibid; p. 207.
77. Meldola, 1915.
78. 'The war and British chemical industry', 1915.
79. 'German industry and the war', 1918.
80. 'The state and industrial research', 1927.
81. Ibid; p. 589.
82. 'Scientific and industrial research', 1917.
83. 'The state and industrial research', p. 589.
84. Letter from W.H. Bragg to A. Smithells, 26 March 1915, cited in Caroe, 1978; p. 138.
85. 'The promotion of textile industries', 1918.
86. 'The physicist in the textile industries', 1923.
87. Letter from W. H. Bragg to A. Smithells, 26 March 1915, cited in Caroe, 1978; p. 138.
88. Ibid.
89. Jenkin, 2011; p. 359.
90. Caroe, 1978; p. 139.
91. Jenkin, 2011; p. 358.

Chapter 3: 'A Keen Young Man'

1. Jenkin, 2011; p. 425; Caroe, 1978; p. 92.
2. Bragg and Caroe, 1962; p. 179.
3. Bernal (1963), p. 2.
4. Astbury, 1942.
5. Hodgkin, 1975.
6. Hodgkin, 1975; p. 452.
7. Ibid.
8. Ibid; p. 453.
9. Lonsdale, 1961.
10. Astbury, 1923.
11. Bernal, 1963; p. 4.

12. Astbury, W.T. (1946–7) 'The science of fibres', *Journal of the Bradford Textile Industry*.
13. Brown, 2005.
14. Bragg, 1920; p. 195. Cited in Caroe, 1978; p. 161.
15. Brown, 2005; pp. 400–402.
16. Hodgkin, 1980; p. 16.
17. Hunter, 2000; p. 193.
18. Brown, 2005; p. 56.
19. Bragg, 1926.
20. Olby, 1994; pp. 26–28.
21. Astbury, 1942; p. 348.
22. Brown, 2005;p. 63.
23. Bernal, 1963; p. 27.
24. Brown, 2005; p. 63.
25. Ibid.
26. Brown, 2005; p. 79.
27. Ibid; p. 80.
28. Anderson, 1988; p.19.
29. Ibid; p. 37.
30. Olby, 1994; p. 42.
31. Anderson, 1988; p. 23.
32. The Development of Research in the Department of Textile Industries. Memorandum written by J. B. Speakman to the Vice-Chancellor, 15 June 1927. University of Leeds, Central Records Office 512.F1/0010.
33. Olby, 1994; p. 43.
34. Letter from W. H. Bragg to A. F. Barker, 19 May 1928. Leeds University Archive. Cited in Olby, 1994; p. 43.
35. Letter from J. B. Speakman to the Vice-Chancellor, 18 September 1917. University of Leeds, Central Records Office 512.F1/0010.

Chapter 4: 'Into the Wilderness'

1. Letter from W. T. Astbury to J. D. Bernal, 13 September 1928. In: John Desmond Bernal: Scientific and Personal Papers. Cambridge University Library, Department of Manuscripts and University Archives GBR/0012/MS Add.8287 J2 (hereinafter referred to as 'JDB papers').
2. Ibid.
3. Olby,1994; p. 45.
4. Ibid; p. 46.
5. Bernal, 1963; p. 26.
6. Preston, 1974; p. 7.
7. Ibid; p. 9.

8. It might seem surprising that Astbury did not carry these studies out using the X-ray spectrometer that his mentor Sir William Bragg had invented while at Leeds and that had helped in earning his Nobel Prize. The main difference between using the X-ray spectrometer and the older photographic method, which Astbury reverted to using, to study diffracted X-rays was that while the spectrometer gave quantitative information, it only did so for X-rays scattered in one particular direction at a time. To record the complete scattering pattern would therefore have required many individual experiments. By contrast, an X-ray photograph showed spots arising from the sum total of scattered X-rays, and so much more information could be obtained from just one single experiment. Professor A. C. T. North, personal communication.

9. A report in preparation for the book *Fifty Years of X-ray Diffraction*, cited in Bernal, 1963; p. 7.

10. Astbury and Woods, 1931a.

11. Astbury and Street, 1932.

12. Astbury, 1933a.

13. Astbury, 1933b.

14. Astbury and Woods, 1933.

15. Astbury and Sisson, 1935.

16. Hunter, 2000; p. 179.

17. Olby, 1994; pp. 11–21.

18. Judson, 1996; p. 22.

19. Morange, 1998; p. 94.

20. Meyer and Mark, 1928. Cited in Olby, 1994; pp. 34–35.

21. Olby, 1994; pp. 35–36.

22. Hunter, 2000; p. 172.

23. Astbury and Street, 1932; p. 96.

24. Astbury and Woods, 1933; p. 334.

25. Bernal, 1961; p. 5.

26. Astbury, 1938.

27. Preston, 1974; p. 11.

28. Astbury, 1952.

29. Although Astbury certainly popularised the phrase 'molecular biology' in the UK with the zeal of an evangelist, the origin of the term has actually been credited to Warren Weaver, who was Director of Natural Sciences at the Rockefeller Institute. Judson, 1996; pp. 52–53.

30. Letter from W. T. Astbury to J. D. Bernal, 16 May 1934. JDB papers; see (n 1).

31. Letter from W. T. Astbury to J. D. Bernal, 23 January 1931. JDB papers; see (n 1).

32. Letter from W. T. Astbury to J. D. Bernal, 26 March 1930. JDB papers; see (n 1).

33. Letter from W. T. Astbury to J. D. Bernal, 23 January 1931. JDB papers; see (n 1).

34. Astbury and Lomax, 1934.

35. Bernal and Crowfoot, 1934.

36. Chibnall, 1964; p. 11.

37. Astbury and Lomax, 1935.

38. Astbury, Dickinson, and Bailey, 1935.

39. Albert Charles Chibnall, Kenneth Bailey, William Thomas Astbury. Production of Filamentary Materials. US Patent 2,358,383; Application 12/3/41 US; 22/10/35 in Great Britain, Australia, Greece, and India.

40. Letter from W. T. Astbury to J. D. Bernal, 16 May 1934. JDB papers; see (n 1).

41. Nor was this effect confined solely to the transformation of hair. In a letter to Bernal, Astbury said that he had been able to induce a perm into the horn of a cow. Yet while Astbury may have given the molecular explanation for how this was done, the actual technique of doing so had already been described by Pliny as being used by cattle thieves to change the appearance of stolen cows so that their original owners could no longer recognise them. W. T. Astbury to J. D. Bernal, 23 January 1931. JDB papers; see (n 1). See also Bernal, 1963; p. 16.

42. 'Secrets of the Perm: Leeds man "lion of the ladies"', *Daily Express.* Astbury Papers MS419 Box A.1 Press clippings, ULSC BL.

43. 'Professor explains the mystery of the curls', *Yorkshire Evening Post*, 15 May 1936; Astbury Papers, MS419 Box A.1 Press clippings, ULSC BL.

44. For an excellent detailed study of Astbury's media work, see Clarke, 2008.

45. 'The technique of the permanent wave', *Daily Mail*, 16 April 1931. Astbury Papers MS419 Box A.1 press clippings book, ULSC BL.

46. 'The permanent wave', *Manchester Guardian*, 23 March 1931. Astbury Papers MS419 Box A.1 press clippings book, ULSC BL.

47. 'The professor says—the permanent wave is a stretch', *Daily Express*, 13 May 1936. Astbury Papers MS419 Box A.1 press clippings book, ULSC BL.

Chapter 5: 'The X-Ray Vatican'

1. Goertzel and Goertzel, 1995; p. 17.

2. Serafini, 1989; p. 55.

3. Serafini, 1989; p. 74.

4. Hager, 1995; p. 214.

5. Olby, 1994; p. 55.

6. Hager, 1995; p. 179.
7. Hunter, 2000; p. 199.
8. Letter from W. T. Astbury to J. D. Bernal, 26 March 1930. JDB papers, see Chapter 4, note 1.
9. *Yorkshire Post*, 12 September 1933. Astbury Papers, MS419 Box A.1 press clippings book, ULSC BL.
10. Olby, 1994; p. 55.
11. Astbury and Dickinson, 1935b.
12. Astbury and Dickinson, 1935a.
13. Astbury and Dickinson, 1936.
14. Astbury and Bell, 1941.
15. Astbury and Dickinson, 1940.
16. Astbury, Bailey, and Rudall, 1943.
17. Crick, 1958.
18. Astbury, 1934.
19. Astbury and Woods, 1932.
20. Astbury, 1955; p. 220.
21. Cushing, 2005; p. 349.
22. Ibid; p. 353.
23. Preston, 1974; p. 15.
24. Astbury and Preston, 1940.
25. Astbury, W. T. (1955) 'Textile fibres and molecular biology', Transcript of a lecture delivered to the International Textile Congress, Brussels, June 1955. Astbury Papers, MS693/45 ULSC BL.
26. MacArthur, 1961.
27. Astbury, 1955; p. 234.
28. Astbury,1952; p. 4.
29. Letter from Warren Weaver to W. T. Astbury, 27 May 1948. Astbury Papers, MS419 Box E.153 ULSC BL.
30. Astbury,1955; p. 220.
31. Astbury and Woods, 1931b; p. 664.
32. Ibid.

Chapter 6: 'A Pile of Pennies'

1. Greenstein, 1943; p. 530.
2. Cohen and Portugal, 1977; p. 78.
3. Judson, 1996; p. 13.
4. Stanley, 1935.
5. Astbury, 1939a.
6. Bawden et al., 1936.

7. Stent, 1966; p. 3. Expanded Edition, 1992. Eds. John Cairns, Gunther S. Stent & James D. Watson, Cold Spring Harbor Press.

8. Moore, 1989; pp. 394–404.

9. Schrödinger, E. (1944) *What is Life?* Cambridge: Cambridge University Press.

10. Letter from W. T. Astbury to J. D. Bernal 27 July 1937. JDB papers.

11. Astbury and Bell, 1938b.

12. Olby, 1994; p. 65.

13. Signer, Caspersson, and Hammarsten, 1938.

14. Lawrence Bragg to W. T. Astbury, 18 October 1937. University of Leeds, Central Records Office Box 178, University Archive.

15. Tutorial File, Bell, F. O. Girton College Library and Archive, University of Cambridge, GCAC 2/4/1/1.

16. Letter from W. T. Astbury to E. Beighton 31 July 1941. Astbury papers, MS419 Box A.10, ULSC BL.

17. Reference from the Headmistress, Haberdashers' Aske's School, West Acton, 18 January 1932. In Tutorial File, Bell, F. O.

18. Tutorial File, Bell, F. O.

19. Lawrence Bragg to W. T. Astbury, 18 October 1937.

20. Ibid.

21. See newspaper cuttings in Astbury Papers, MS419 A.1 ULSC BL. Also *Yorkshire Evening News*, 23 March 1939.

22. In his chronicle of the life of the famous eighteenth-century polymath and writer Dr Samuel Johnson, James Boswell records that on Sunday, 31 July 1763, he told Johnson that he had 'been that morning at a meeting of the people called Quakers, where I had heard a woman preach', to which Johnson made the infamous reply—'Sir, a woman's preaching is like a dog's walking on its hinder legs. It is not done well; but you are surprised to find it done at all' Boswell, J. (1791/1986) *The life of Samuel Johnson*. London: Penguin Books, p. 116). In her biography of Rosalind Franklin, *Dark Lady of DNA*, Brenda Maddox uses this same quotation from Johnson to describe Rosalind Franklin's treatment (Maddox, 2002; pp. 133–134).

23. Letter from W. T. Astbury to W. L. Bragg, 5 November 1943. Private papers of Professor A. C. T. North. Cited with kind permission of Professor A. C. T. North.

24. Olby, 1994; p. 65.

25. Letter from Florence Sawyer to Robert Olby, 1967. Cited in Olby, 1994; p. 67.

26. Astbury and Bell, 1938a; p. 113.

27. Astbury, 1952; p. 39.

28. Olby, 1994; p. 70.

29. Astbury and Bell, 1938b; p. 114.

30. Judson, 1996; p. 13.
31. Astbury, 1939a; p. 49.
32. Astbury, 1939b; p. 123.
33. According to some sources, the experimental evidence for Levene's tetranucleotide hypothesis was actually quite scarce. Rollin Hotchkiss, a US biochemist who worked with the clinician Oswald Avery on the role of nucleic acids in pneumococcus, examined a monograph published in 1931 by P. A. Levene and L. W. Bass that was universally cited as being the classical statement of the tetranucleotide theory: Levene, P. A. and Bass, L. W. (1931) *Nucleic acids*. New York: Chemical Catalog Company. Hotchkiss found that the correlation of experimentally determined percentages of nucleotides with the theoretically predicted values was, at best, only very approximate. Hotchkiss explained this disagreement as being due to the limitations of the chemical methods that Levene would have employed, but said that, as a result, he felt that 'we need not take the values too seriously'. Furthermore, he claimed that several people working on nucleic acids in the 1940s 'all stated that we didn't feel obliged to take the tetranucleotide calculations seriously'. See Judson, 1996; p. 13. See also Judson, 1980; p. 406.
34. Astbury and Bell, 1938b; p. 114.
35. Schultz, 1941.
36. Bell, F. O. (1939) *X-ray and related studies of the structure of the proteins and nucleic acids*. PhD thesis, University of Leeds, Brotherton Library.
37. Maddox, 2002; p. 202.
38. Watson, 1968; p. 54.
39. Astbury and Bell, 1938b; p. 114.
40. Astbury and Bell, 1938a.
41. Astbury and Bell, 1938b); p. 114.
42. *New York Times*, 30 August 1939. Astbury Papers, MS419 Box A.1 press clippings book, ULSC BL.
43. Seventh International Genetical Congress (1939) *Nature*, 144, p. 496.
44. Seventh International Congress of Genetics (1939) *Science*, 89, pp. 528–529.
45. The International Congress of Genetics (1939) *Science*, 90, p. 228.
46. Seventh International Genetical Congress (1939) *Nature*, 144, p. 496.
47. Seventh International Genetical Congress (1939) *Nature*, 144, p. 496.
48. Letter from W. T. Astbury to A. Melland, 14 June 1940. Astbury Papers, MS419 Box E.119, ULSC BL.
49. Letter from W. T. Astbury to the Vice-Chancellor, 20 May 1940. Astbury Papers, MS419 Box B.18, ULSC BL.
50. Letter from W. T. Astbury to J.D. Bernal, 4 April 1941. Astbury Papers, MS419 Box E.13, ULSC BL.

51. Letter from W. T. Astbury to I. Fankuchen, 24 April 1942. Box 1 Folder 6, Isidor Fankuchen Papers, 1933–64. American Institute of Physics, Niels Bohr Library and Archives. Cited in Olby, 1994; p. 119.

52. Letter from W. T. Astbury to C. P. Snow, Secretary of the Central Register, 5 May 1941, Astbury Papers, MS419 Box B.1, ULSC BL.

53. Letter from W. T. Astbury to A. E. Wheeler (Registrar, University of Leeds), 18 December 1941. Astbury Papers MS419 Box B.4, ULSC BL.

54. Letter from A. E. Wheeler to F. O. Bell, 9 June 1942. University of Leeds, Central Records Office Box 178, University Archive.

55. Letter from F. O. Bell to A. E. Wheeler, 4 January 1943. University of Leeds, Central Records Office Box 178, University Archive.

56. Girton College Annual Review, 2002. Girton College Archive Ref. GCCP 10/1/1. Girton College, University of Cambridge.

57. Lederberg, 2000.

58. Watson et al., 1987; p. 69.

59. Letter from W. T. Astbury to F. B. Hanson, 19 October 1944. Astbury Papers, MS419 Box E.152 ULSC BL.

Chapter 7: 'Avery's Bombshell'

1. Dubos, 1976; p. 10.

2. Dubos, 1976; p. 71.

3. Dubos, 1976; p. 161.

4. Dubos, 1976; p. 70.

5. Dowling, 1977; p. 230. Cited in Amsterdamska, 1993; p. 7.

6. Cole, R. (1915) 'Report of the Director of the Hospital to the Corporation of the Rockefeller Institute for Medical Research, 16 January 1915'. Scientific Reports of the Laboratories to the Board of Scientific Directors, Record Group 439, Rockefeller University Archives, Rockefeller Archive Center, Sleepy Hollow, New York. In: *The Oswald T. Avery Collection*. Profiles in Science, National Library of Medicine; <http://profiles.nlm.nih.gov/ps/access/CCAAER.pdf>.

7. Cole, R. (1916) 'Report of the Director of the Hospital to the Corporation of the Rockefeller Institute for Medical Research, 10 October 1916'. Scientific Reports of the Laboratories to the Board of Scientific Directors, Record Group 439, Rockefeller University Archives, Rockefeller Archive Center, Sleepy Hollow, New York. In: *The Oswald T. Avery Collection*. Profiles in Science, National Library of Medicine; <http://profiles.nlm.nih.gov/ps/access/CCAAEY.pdf >.

8. Dubos, 1976; p. 104.

9. That bacteria could be identified as R or S forms depending on whether they possessed an outer capsule was first observed in 1921 by the British bacteriologist J. A. Arkwright, who was working on Shiga dysentery bacteria. Arkwright also observed that only the S forms of the bacteria were virulent.

10. Griffith, F. (1923) 'The influence of immune serum on the biological properties of pneumococci. Reports on Public Health and Medical Subjects', in Bacteriological Studies,18, London: HMSO, pp. 1–13.

11. Dubos, 1976; p. 132.

12. Griffith, 1928.

13. Dubos, 1976; p. 133.

14. Dawson and Sia, 1931a.

15. Dawson and Sia, 1931b.

16. Alloway, 1932.

17. Avery et al., 1944; p. 155.

18. Cobourn, 1969; p. 630.

19. Letter from Avery to his brother Roy, 26 May 1943. Cited in Dubos, 1976; p. 219.

20. Stent, 1972.

21. Maddox, 2002; p. 120.

22. Wyatt, 1972.

23. In his book *A History of Molecular Biology*, Michell Morange says that Avery has often been cast as 'molecular biology's equivalent of Mendel' (Morange, 1998; p. 30). At first sight, the monk Gregor Mendel does seem to be a classic example of the 'neglected genius' whose work goes unrecognised as a result of being 'ahead of its time'. Working in his monastery garden in Brno in the 1860s, Mendel made careful studies of the numerical ratios in which specific physical traits such as seed shape and colour occur in the progeny of self-fertilised plants. From this work emerged the idea that biological traits are determined by discrete entities within the gametes. As a result of this work, Mendel is attributed the status of the founding father of genetics and heredity by most biology textbooks. Mendel himself would most likely have been surprised to find his work revered in this way. But this is an interpretation that has long been challenged by historians of science; see Olby, 1979, who points out, in the introduction to his paper Mendel declared that his aim was to discover a law that explained how new species are produced by the hybridisation of pre-existing ones. What to twenty-first-century eyes appears to be a pioneering quantitative study in heredity was to Mendel ultimately a rather ambiguous attempt to discover the mechanism by which new species evolve.

24. Hotchkiss, 1955.
25. Dubos, 1976; p. 146.
26. Hershey and Chase, 1952.
27. Olby, 1994; p. 202.
28. Lederberg, 1972.
29. Olby, 1972.
30. Boivin, 1947.
31. Alexander and Leidy, 1951a.
32. Alexander and Leidy, 1951b.
33. Alexander and Leidy 1953.
34. Leidy, Hahn, and Alexander, 1953.
35. Letter from Avery to his brother Roy, 26 May 1943; see (n 20).
36. Watson, 1968; pp. 13–14 and 23.
37. Gulland, Barker, and Jordan, 1945.
38. An entry in Astbury's travel diary reads: 'Then talked about possible orientation in the polysaccharide capsule of pneumococcus. Sketched out plan of campaign with Goebel (c.f. Valonia). Introduced to polysaccharide assistant, Dr. R. Hotchkiss & then to the immunological head, Dr. O.T. Avery. Discussed problem with him too. Saw fibrous polysaccharides from pneumococci. Goebel will send samples and also dead cocci for X–rays. Most exciting!'. Astbury (1934), p. 18. Astbury Papers MS419 Box A.4, ULSC BL.
39. Letter from W. T. Astbury to W. T. J. Morgan, 8 August 1941. Astbury Papers, MS419 Box E.122, ULSC BL.
40. Letter from W. T. Astbury to W. T. J. Morgan, 18 July 1942. Astbury Papers, MS419 Box E.122, ULSC BL.
41. Letter from W. T. Astbury to W. T. J. Morgan, 12 October 1942. Astbury Papers, MS419 Box E.122, ULSC BL.
42. Morgan, 1944.
43. Astbury, 1945b.
44. Letter from W. T. Astbury to F. B. Hanson, 19 October 1944. Astbury Papers MS419 Box E.152, ULSC BL.
45. Stern 1947)
46. Letter from W. T. Astbury to F. B. Hanson, 19 October 1944. Astbury Papers MS419 Box E.152, ULSC BL.
47. Letter from W. T. Astbury to O. T. Avery, 18 January 1945. Astbury Papers MS419 Box E.152, ULSC BL.
48. Letter from Avery's colleague MacLyn McCarty to Joshua Lederberg, 9 June 1981. Profiles in Science, National Library of Medicine, The Oswald Avery Collection; <http://profiles.nlm.nih.gov/ps/access/CCAAEA.pdf >.
49. Letter from W. T. Astbury to F. B. Hanson, 19 October 1944. Astbury Papers MS419 Box E.152, ULSC BL.

Chapter 8: 'Nunc Dimittis'

1. Letter from W. T. Astbury to Vice-Chancellor, University of Leeds, 6 February 1945. Astbury Papers MS419 Box B.18, ULSC BL.
2. Letter from W. T. Astbury to F. B. Hanson, 14 February 1945. Astbury Papers, MS419 Box E.152, ULSC BL.
3. Letter from W. T. Astbury to F. B. Hanson, 19 October 1944. Astbury Papers, ULSC Box E.152, ULSC BL.
4. The prayer refers to a passage in the gospel of Luke, in which the infant Christ is brought to the temple by his parents. Sitting in the temple is an old man, Simeon, who has spent all his life watching and waiting to see the arrival of the Messiah. On seeing the boy, Simeon knows that, having finally witnessed the moment for which he has waited all his life, he can at last depart this world a fulfilled man, and he joyfully declares, 'Lord, now lettest thou thy servant depart in peace'. Luke 2, v. 29.
5. Item 3648, Senate Records, 2 May 1945, University of Leeds Central Records Office.
6. Item 3697, Committee on Proposed Department of Biomolecular Structure, Senate Records, 6 June 1945, University of Leeds, Senate Records Office.
7. Letter from J. A. Priestley, Professor of Botany, to W. T. Astbury, 29 June 1944. Astbury Papers MS419 Box B.14, ULSC BL.
8. Astbury, W. T. (1961) 'Molecular biology or ultrastructural biology?', *Nature*, 190, p. 1124.
9. Waddington, 1961a.
10. Waddington, 1961b.
11. Letter from W. T. Astbury to Warren Weaver, 11 January 1948. Astbury Papers MS419 Box. E.153, ULSC BL.
12. Letter from W. T. Astbury to Edward Mellanby, 8 December 1945. Astbury Papers MS419 Box. G.13, ULSC BL.
13. Letter from E. Mellanby to W. T. Astbury, 21 March 1946. Astbury Papers MS419 Box G.13, ULSC BL.
14. Letter from W. T. Astbury to E. Mellanby, 22 March 1946. Astbury Papers MS419 Box G.13, ULSC BL.
15. Letter from H. Himsworth to R. Olby, 17 May 1968. Cited in Olby, 1994; p. 327.
16. Letter from W. T. Astbury to Miller, 23 March 1946. Astbury Papers MS419 Box E.153, ULSC BL.
17. Wilkins, 1987.
18. Ibid; p. 505.
19. Wilkins, 2003; p. 97.
20. Ibid, p. 98.

21. Letter from W. T. Astbury to J. T. Randall, 16 July 1947. Astbury Papers MS419 Box E.147, ULSC BL.
22. W. T. Astbury, 5 July 1947. Astbury Papers, MS419 Box E.44, ULSC BL.
23. Davies, 1990; p. 613.
24. Preston, 1974; p. 19.
25. Hager, 1995; p. 215.
26. Letter from W. T. Astbury to Linus Pauling, 8 November 1943. Astbury Papers, MS419 Box E.135, ULSC BL.
27. Letter from Linus Pauling to W. T. Astbury, 6 December 1943. Astbury Papers, MS419 Box E.135, ULSC BL.
28. Letter from W. T. Astbury to Linus Pauling, 12 February 1945. Astbury Papers, MS419 Box E.135, ULSC BL.
29. Hager, 1995; p. 372.
30. Goertzel and Goertzel, 1995; p. 94.
31. Judson, 1996; p. 66.
32. Hager, 1995; p. 373.
33. Ibid; p. 414.
34. Olby, 1994; p. 144.
35. Hager, 1995; p. 378.
36. Ferry, 2008; p. 146.
37. Judson, 1996, p. 69.
38. Ibid; pp. 555–556.
39. Letter from W. T. Astbury to Dr C. G. Kay Sharp, 21 September 1946. Astbury Papers MS419 Box B.19, ULSC BL.
40. Letter from Clerk of Fabric Office, University of Leeds to W. T. Astbury, 15 August 1946. Astbury Papers MS419 Box B.20, ULSC BL.
41. Letter from Clerk of Fabric Office, University of Leeds to W. T. Astbury, 28 September 1948. Astbury Papers MS419 Box B.20, ULSC BL.
42. Letter from W. T. Astbury to Mr G. Wilson, 23 February 1950. Astbury Papers MS419 Box B.10, ULSC BL.
43. Letter from A. L. Knighton, Surveyor of Fabric for University of Leeds, to W. T. Astbury, 16 February 1952. Astbury Papers MS419 Box B.20, ULSC BL.
44. Letter from W. T. Astbury to A. L. Knighton, 12 February 1952. Astbury Papers MS419 Box B.20, ULSC BL.
45. Letter from W. T. Astbury to A. L. Knighton, 6 November 1953. Astbury Papers MS419 Box B.20, ULSC BL.
46. Letter from W. T. Astbury to Planning Engineer, 21 May 1952. Astbury Papers MS419 Box B.20, ULSC BL.
47. Letter from W. T. Astbury to Mrs M. Kelly, Superintendent of Cleaners, University of Leeds, 18 June 1953. Astbury Papers MS 419 Box B.16, ULSC BL.

48. Letter from W. T. Astbury to Brown, 5 October 1951. Astbury Papers MS419 Box B.11, ULSC BL.
49. Letter from W. T. Astbury to Mr Cullen, 7 November 1950. Astbury Papers MS419 Box B.14, ULSC BL.
50. Letter from W. T. Astbury to G. D. Preston, 29 September 1941. Astbury Papers MS419 Box E.143, ULSC BL.
51. Letter from W. T. Astbury to F. B. Hanson, 6 August 1943. Astbury Papers MS419 Box E.152, ULSC BL.
52. Letter from W. T. Astbury to Flt Lt H. S. Hoff, 9 August 1943. Astbury Papers MS419 Box B.30, ULSC BL.
53. Letter from P. Culpam to W. T. Astbury, 26 April 1944. Astbury Papers MS419 Box B.28, ULSC BL.
54. Letter from W. T. Astbury to E. M. Gilligan, 12 October 1944. Astbury Papers MS419 Box B.28, ULSC BL.
55. Letter from W. T. Astbury to Happold, 10 March 1945. Astbury Papers MS 419 Box B.14, ULSC BL.
56. Kellgren et al., 1951.
57. Passey, Dmochowski, Reed, and Astbury, 1950.
58. Ibid; 1950.
59. Passey et al., 1951.
60. Letter from W. T. Astbury to the Vice-Chancellor, University of Leeds, 16 June 1955. Astbury Papers MS419 Box B.18, ULSC BL.
61. *Yorkshire Evening Post*, 22 October 1953; Astbury Papers, MS419 Box A.3 press clippings book, ULSC BL.
62. Maddox, 2002; p.141.
63. Randall, J. T. (1946) Programme of Biophysics Research to be carried out by Professor J. T. Randall, F.R.S., in King's College, University of London. 22 July 1946. Churchill Archives Centre, University of Cambridge, The Papers of John Randall, GBR/0014/RNDL 2/2/1.
64. Ibid.
65. Wilkins, 2003; p. 85.
66. Ibid; pp. 83–85.
67. Ibid; pp. 83–85.
68. Ibid; p. 107.
69. Ibid; p. 115.
70. Ibid; p. 116.
71. Ibid; p. 137.
72. Ibid; p. 90.
73. Ibid; p. 117.
74. Ibid; p. 122.
75. Report on the Naples Conference 1951, in a letter from W. T. Astbury to R. Dohrn, 7 September 1951. Astbury Papers, MS419 Box F.4, ULSC BL.

76. Wilkins, 2003; p. 137.

77. Ibid; p. 140.

78. Ibid; p. 101.

79. Ibid; p. 103.

80. Ibid; p. 106.

81. Maddox, 2002; p. 114. See also Wilkins, 2003; p. 144.

82. Maddox, 2002; p. 130. See also Wilkins, 2003; p. 144.

83. Maddox, 2002; pp. 128–129.

84. Letter from W. T. Astbury to J. T. Randall, 18 April 1951. Astbury Papers, MS419 Box E.147 ULSC BL.

85. Letter from J. T. Randall to W. T. Astbury, 19 April 1951. Astbury Papers, MS419 Box E.147, ULSC BL.

86. Letter from W. T. Astbury to J. T. Randall, 11 May 1951. Astbury Papers, MS419 Box E.147, ULSC BL.

87. Gordon, 1996; p. 459.

88. Gordon, H. (2004) 'Synge, Richard Laurence Millington (1914–1994)', in *Oxford Dictionary of National Biography*. Oxford: Oxford University Press, pp. 621–623. Online edition, October 2006; <http://www.oxforddnb.com/view/article/55773>.

89. Ibid; p. 460.

90. Lovelock, 2004; p. 160.

91. Martin, A. J. and Synge, R. L. (1941) 'Some applications of periodic acid to the study of the hydroxyamino-acids of protein hydrolysates: the liberation of acetaldehyde and higher aldehydes by periodic acid. 2. Detection and isolation of formaldehyde liberated by periodic acid. 3. Ammonia split from hydroxyamino-acids by periodic acid. 4. The hydroxyamino-acid fraction of wool. 5. Hydroxylysine. With an Appendix by Florence O. Bell, Textile Physics Laboratory, University of Leeds', *Biochemical Journal*, 35, pp. 294–314.

92. Letter from W. T. Astbury to K. Bailey, 1 November 1941. Astbury Papers MS419 Box E.6, ULSC BL.

93. The London Times compared Sanger's success at sequencing the amino acids in insulin with the achievement of the athlete Sir Roger Bannister in running the 4 minute mile. 'Dr. Sanger Awarded Nobel Prize for Chemistry,' The Times, October 29, 1958, 8.

94. Sanger 1988, 9; Sanger 1952, 2.

95. For much more detail about the story of Martin and Synge and the important legacy of their work, see Chapters 7–9 in 'Insulin – the Crooked Timber: A History From Thick Brown Muck to Wall Street Gold' by K.T. Hall (Oxford University Press, 2022).

Chapter 9: 'One Grand Leap' . . . Too Far

1. Judson, 1996; p. 634.

2. Hunter, 2000; p. 290.
3. Chargaff, 1978; p. 47.
4. Chargaff, 1978; p. 83.
5. Gulland, Barker, and Jordan, 1945.
6. Gulland, 1947.
7. Consden, Gordon, and Martin, 1944.
8. Olby, 1994; p. 436.
9. Chargaff and Zamenhof, 1948.
10. Vischer and Chargaff, 1947; Chargaff and Vischer, 1948b.
11. Vischer, Zamenhof, and Chargaff, 1949; Chargaff and Vischer, 1948a.
12. Chargaff et al., 1949.
13. Vischer, Zamenhof, and Chargaff, 1949.
14. Chargaff, Zamenhof, and Green, 1950.
15. Chargaff, 1950; p. 202. It is also in this paper that Chargaff cites the work of the Torridon scientists Consden, Gordon, and Martin for their development of the chromatography method that he applied to nucleic acids.
16. Judson, 1996; p. 93.
17. Astbury, 1947; p. 68.
18. Letter from W. T. Astbury to E. Chargaff, 14 March 1951. Astbury Papers MS419 Box E.28, ULSC BL.
19. Chargaff, 1978; p. 99.
20. Letter from E. Chargaff to W. T. Astbury, 19 March 1951. Astbury Papers MS419 Box E.28, ULSC BL.
21. Letter from A. E. Wheeler, Registrar, University of Leeds, to W. T. Astbury, 13 May 1939. Astbury Papers MS419 Box B.4, ULSC BL.
22. Letter from W. T. Astbury to Secretary, Ministry of Labour and National Service, 29 January 1941. Astbury Papers MS419 Box B.1, ULSC BL.
23. Letter from E. Beighton to W. T. Astbury, 29 July 1941. Astbury Papers MS419 Box B.1, ULSC BL.
24. Letter from E. Beighton to W. T. Astbury, 29 November 1945. Astbury Papers MS419 Box B.1, ULSC BL.
25. Ibid.
26. Chargaff, 1978; p. 99.
27. Letter from E. Chargaff to W. T. Astbury, 19 March 1951. Astbury Papers MS419 Box E.28, ULSC BL.
28. Olby (1994), p. 379.
29. Letter from R. Olby to the University of Leeds Special Collections, 20 July1992. Astbury Papers MS419 C.7, ULSC BL.
30. Davies, 1990.
31. Ibid; p. 615.

32. Preston, 1974; p. 18.
33. Ibid; p.19.
34. Letter from W. T. Astbury to Sir Charles Martin, 29 October 1945. Astbury Papers MS419 Box E.115, ULSC BL.
35. Recollection of Mr W. Astbury (grandson); personal communication.
36. Bailey, 1961; p. xi.
37. Letter from W. T. Astbury to Jerome Alexander, 31 October 1950. Astbury Papers MS419 Box A.6, ULSC BL.
38. Letter from W. T. Astbury to Linus Pauling, 18 April 1952. Astbury Papers MS419 Box E.135, ULSC BL.
39. Preston, 1974; p. 19.
40. Bailey, 1961; p. xi.
41. Preston, 1974; p. 19.
42. Astbury, W. T. (1955) 'Textile Fibres and Molecular Biology', Transcript of a lecture delivered to the International Textile Congress, Brussels, June 1955. Astbury Papers, MS693/45 ULSC BL.
43. Astbury, 1960; p. 525.
44. The story of how Mozart's hair came to be in the possession of the University of Leeds and a test subject for Astbury's X-ray analysis could probably fill a book in itself. In 1829, the composer Vincent Novello and his wife Mary set out on a trip to Europe with the aim of presenting Mozart's elderly sister with a sum of money that they had collected and also to gather material for a projected life of Mozart. Throughout their trip, Vincent and his wife kept travel diaries. These were eventually tracked down after Major Edward Croft Murray, a British army officer, was quartered in the Villa Novello outside Genoa in 1944, where he found a book that referred to the diaries. The diaries were published as a translation in 1955 by Rosemary Hughes as *A Mozart Pilgrimage* and described Vincent and Mary's visit to Mozart's widow Constanze in Salzburg. According to the diaries, during one of these visits, Constanze presented Vincent and Mary with a lock of her husband's hair. In 1950, the University of Leeds was presented with the travel diaries, together with other papers and the lock of hair by Novello's Italian descendants. Although their connection with the University of Leeds is not entirely clear, it is thought that since the University of Leeds was known to be the 'most serious, active institutional collector of 19th century rare books' that it was the ideal home for the Novello collection. Professor A. C. T. North, University of Leeds, personal communication, 2011.
45. Letter from W. T. Astbury to Warren Weaver, 25 May 1948. Astbury Papers MS419 Box E.153, ULSC BL.
46. Astbury, W. T. (1953) 'How to Swim with a Molecule for a Tail', Lecture delivered at Wayne State University School of Medicine, 10 October 1953. Recording provided by Jim Garretts. Cited with kind permission of Professor Laszlo Lorand.

47. Letter from W. T. Astbury to A. V. Hill, 17 November 1948. Churchill Archives Centre, Churchill College, Cambridge, AVHL II 4/5.
48. Astbury, W. T. (1957) 'BBC Lecture: The Chemical Basis of Life', Recorded 31 January 1957; transmitted 7 February 1957. Transcript in Astbury Papers 693/55 ULSC BL.
49. Letter from W. T. Astbury to A. V. Hill, 17 October 1951. Churchill Archives Centre, Churchill College, Cambridge, AVHL II 4/5.
50. Astbury and Weibull, 1949.
51. Astbury, 1951.
52. Astbury, 1952; p.28.
53. Abir-Am, 1982.
54. Letter from W. T. Astbury to J. T. Randall, 26 September 1945. Astbury Papers MS419 Box E.147, ULSC BL.
55. For a robust response to the charge that Astbury was not really interested in biology, see Olby, 1984.
56. Davies, 1990; p. 614.
57. Recollection of Mrs S. Sanderson and Ms Gemma Sanderson (daughter and granddaughter of Elwyn Beighton); personal communication.
58. Beighton, E. (1952) *X-ray studies of the structure of bacterial flagella*. PhD thesis, University of Leeds. Library shelfmark THESES S/BEI, pp. 29–30.
59. Astbury, 1952; p. 31.
60. Ibid.
61. Ibid.
62. Wilkins, 2002; p. 119–121.
63. Cochran, Crick, and Vand, 1952.
64. Olby, 1984; p. 246. That Astbury had not kept up to date with the latest developments in the theory of X-ray diffraction showing that a helical structure would give a cross pattern has also been suggested by Professor A. C. T. North. Mansell Davies cites a letter written to him by Professor North in May 1989; Davies, 1990; p. 614. See also Olby, 1994; p. 436.
65. Vischer, Zamenhof, and Chargaff, 1949.
66. Chargaff, Zamenhof, and Green, 1950.
67. Judson, 1996; p. 633.
68. Ibid.
69. Olby, 1994, p. 388.
70. Gulland, Jordan, and Taylor, 1947.
71. Creeth, Gulland, and Jordan, 1947.
72. Judson, 1996; p. 144.
73. Gulland, Jordan, and Taylor,1947.
74. Letter from J. M. Gulland to W. T. Astbury, 19 March 1946. Astbury Papers MS419 Box E.60, ULSC BL.
75. Manchester, 1995.

76. Actually, in what may at first sight seem like a tantalising moment of pre-science, Astbury did consider some kind of spiral pattern for the nucleic acid chain. Back in 1939, when he and Bell had first suggested that nucleic acid and protein might work as a hybrid nucleoprotein, he had specu-lated that 'Possibly there is a spiralized column of nucleotides about which the protein chains entwine themselves'. But the context in which he was thinking was very different to that of Watson and Crick, for whom the sig-nificance of the helix was that it enabled pairing between bases on opposite strands; Astbury, 1939a; p. 49.

77. Astbury, 1938; p. 380.

78. Letter from W. T. Astbury to F. B. Hanson, 19 October 1944. Astbury Papers MS419 Box E.152, ULSC BL.

79. Olby, 1994; p. 379.

80. One of Astbury's only comments on Watson and Crick's work was the following reference in a lecture given in 1960 on biological fibres: 'The supreme illustration with which to conclude is obviously that of the nucleic acids. When Watson and Crick inferred that deoxyribonucleic acid (DNA) is a two-strand helix in which certain nucleotides are always matched together, and it was shown to be so by X-rays, then was inau-gurated with all its momentous consequences the science of "molecular genetics". We know today that the sequence of nucleotides in long chains of DNA carries the ultimate code of instructions for biosynthesis, and what more could one ask of a fibre?'; Astbury, 1960; p. 525.

81. Olby, 2003.

82. Ridley, 2006; p. 73.

83. Commoner, 1968.

84. Astbury, 1939a; p. 51.

85. Astbury, 1958; p. 94: 'Deoxyribonucleic acid and ribonucleic acid have lately come to be recognized as the "master race" of biological high polymers, in that apparently they alone can both transmit messages of heredity and afterwards dominate the business of protein synthesis.'

86. Astbury, 1952; p. 42.

87. Ibid; p. 39.

88. Olby, 1994; p. 70.

89. Astbury, 1947; p. 70.

90. Astbury, 1952; p. 41.

91. Astbury, 1939b; p. 129.

92. Astbury, 1952; p. 38.

93. Fox Keller, 2000. It is also becoming evident that the precise way in which the genetic information contained in DNA is expressed can be significantly influenced by its particular cellular environment. For more on this new and exciting science of epigenetic, see Carey, N. (2012) *The epigenetics revolution:*

how modern biology is rewriting our understanding of genetics, disease and inheritance. New York: Columbia University Press.

94. Rose, 1997; pp. 125–126 and 171. Cited in Midgley, 2001; p. 7.
95. Crick, cited in Olby, 2009; 467, note 26.
96. See the article by Professor Matthew Cobb, 'Happy 100th Birthday, Francis Crick (1916–2004)', in which he writes: 'Watson's own description of the discovery of the structure of DNA did not contain any striking new revelations, with one exception. He finally admitted that when he wrote in *The Double Helix* that Crick strode into the Eagle pub and proclaimed, "We have discovered the secret of life," this was not true. Watson said he made it up, for dramatic effect. Crick always denied saying any such thing, and historians have long known that *The Double Helix* cannot be taken as an entirely reliable source.' Online at https://whyevolutionistrue.com/2016/06/08/happy-100th-birthday-francis-crick-1916-2004/.
97. Crick, 1988; p. 77.
98. Wilkins, 2003; p. 137.
99. Olby, 2009; p. 129.

Chapter 10: 'The Road Not Taken . . .'

1. 'The Road Not Taken', from *Mountain Interval*, 1920, a poem by Robert Frost (1874–1963).
2. Pauling, L. (1948) 'Molecular Architecture and the Processes of Life', Twenty-First Sir Jesse Boot Foundation Lecture. Quoted in Olby, 1994; p. 120.
3. Olby, 1994; p. 366.
4. Watson, 1968; pp. 145–146.
5. Serafini, 1989; p. 155.
6. Ibid.
7. Olby, 2009; p. 158.
8. Maddox, 2002; p. 191.
9. Hager, 1995; p. 397.
10. Letter from J. T. Randall to Linus Pauling, 28 August 1951. Oregon State University Special Collections, Ava Helen and Linus Pauling Papers 1873–2011, 11. Science Papers 1923–1994; Nucleic Acid Papers 1951–1963, Box #9.001 Folder #1.2. Online at <http://osulibrary.oregonstate.edu/specialcollections/coll/pauling/calendar/1951/08/28.html>. Cited by courtesy of Ava Helen and Linus Pauling Papers 1873–2011, Oregon State University Special Collections.
11. Maddox, 2002; p. 152.
12. Letter from W. T. Astbury to L. Pauling, 20 September 1951. Astbury Papers MS419 Box E.131, ULSC BL.

13. Letter from W. T. Astbury to L. Pauling, April 1952. Astbury Papers MS419 Box E.131, ULSC BL.
14. Letter from L. Pauling to Astbury, 25 April 1952. Astbury Papers MS419 Box E.131, ULSC BL.
15. Hager, 1995; p. 380.
16. Goertzel, 1995; pp. 118–119.
17. Hager, 1995; p. 383.
18. Goertzel, 1995; p. 123.
19. Hager, 1995; p. 401.
20. Letter from W. T. Astbury to L. Pauling, 13 May 1952. Astbury Papers MS419 Box E.131, ULSC BL.
21. Albert Einstein to Dean Acheson, Secretary of State, 21 May 1952. Oregon State University Special Collections, Ava Helen and Linus Pauling Papers 1873–2011. Correspondence Box #107, Folder #1. Cited in Serafini; 1989; p. 292.
22. Serafini, 1989; p. 151.
23. Olby, 1994; p. 378.
24. Goertzel, 1995; p. 130.
25. Hager, 1995; p. 414.
26. Letter from Linus Pauling to J. T. Randall, 31 December 1952. Oregon State University Special Collections. Ava Helen and Linus Pauling Papers 1873–2011. 11. Science Papers 1923–1994; Nucleic Acid Papers 1951–1963, Box #9.001 Folder #1.18. Online at <http://osulibrary.oregonstate.edu/specialcollections/coll/pauling/calendar/1952/12/31.html>. Cited by courtesy Ava Helen and Linus Pauling Papers, Oregon State University Special Collections, Box 6.
27. Pauling and Corey, 1953.
28. Hager, 1995; p. 430.
29. Ibid.
30. Pauling, cited in Maddox, 2002; p. 192.
31. Ridley, 2006; p. 74.
32. Letter from Linus Pauling to J. T. Randall, 31 December 1952; see note 26.
33. Hager, 1995; p. 415.
34. Letter from W. T. Astbury to Linus Pauling, 6 October 1952. Oregon State University Special Collections. Ava Helen and Linus Pauling Papers 1873–2011. 01 Correspondence 1919–2000; Box 6.18.
35. Hager, 1995; p. 415.
36. *Leeds Mercury*, 3 August 1955. Astbury Papers, MS419 Box A.3 Press clippings, ULSC BL.
37. *Yorkshire Post* newspaper, 21 November 1955. Astbury Papers, MS419 Box A.3 Press clippings, ULSC BL.

38. Professor M. Losowsky, personal communication. See 'Appendix B: 'Laszlo Lorand—The Biochemist Who Came in From the Cold.'
39. Letter from W. T. Astbury to the Editor, published in the *Yorkshire Post*, 18 November 1955; Astbury Papers, MS419 Box A.3 Press clippings, ULSC BL.
40. Death certificate for W. T. Astbury. Private papers of Mr W. Astbury (grandson). Cited with kind permission of Mr W. Astbury.
41. Dr K. D. Parker, personal communication.
42. Preston, 1974; p. 18.
43. Ibid; p. 19.
44. Goertzel, 1995; p. 90.
45. Pauling et al., 1949.
46. Ferry, 2008, chapter 11, 'Health and Disease'.
47. Ferry, 2008; p. 234.
48. Uversky and Fink, 2007.
49. Goldschmidt et al., 2010.

Chapter 11: The Man in the Monkeynut Coat

1. 'Crack out the liquid nitrogen, dumplings . . . we're on our way'. Cartoon in the *Boston Globe,* 9 February 1977. Document 5.9 in Watson and Tooze, 1981; p. 115.
2. Morange, 1998; p. 192.
3. Smith and Wilcox, 1970.
4. Kelly and Smith, 1970.
5. Konforti, 2000.
6. Jackson, Symonds, and Berg, 1972.
7. Morange, 1998; p. 190.
8. Berg et al., 1974.
9. Rogers, M. (1975) 'The Pandora's Box Congress', Rolling Stone, 19 June 1975. Document 2.1 cited in Watson and Tooze, 1981; p. 28.
10. Morange, 1998; pp. 187–194.
11. Ibid; p. 191.
12. *San Francisco Chronicle*, 4 April 1977. Document 6.11 cited in Watson and Tooze, 1981; p. 165.
13. 'City blocks DNA research', The Washington Post, 9 July 1976. 'Document 5.4 cited in Watson and Tooze, 1981; p. 100.
14. Redfearn, 1980.
15. Dickson, 1980.
16. Newmark, 1983.

17. This idea of molecular biology having an alternative origin in 1930s and 1940s West Yorkshire was first suggested by Professor Greg Radick at the University of Leeds, personal communication.

18. *Yorkshire Evening Post*, 18 January 1944. Astbury Papers MS419 Box A.1 Press clippings, ULSC BL. Also, Leeds Central Library records.

19. 'X-Rays and the Study of Textile Fibres (Science Lifts the Veil)', 2 March 1942; 'Textile Fibres from Peanuts', Calling the West Indies—Science and Agriculture Series, 15 January 1943; 'Textiles from Monkey Nuts', Science Notebook, 19 July 1943; 'Fibres—Textiles from Nuts, Beans and Milk', 14 January 1944. Cited in Clarke, 2008. Cited courtesy of BBC Written Archive Centre. There is also correspondence pertaining to these broadcasts in Astbury Papers MS419 Box D.4, ULSC BL.

20. 'X-rays and the study of textile fibres' (1942).

21. Ibid.

22. Ibid.

23. Actually, Astbury's speculation about being able to refold virus proteins into fibres for use in textiles was not entirely fanciful. In 2010, scientists at Imperial College London in collaboration with fashion designers from Central Saint Martin's College of Art and Design, announced the creation of artificial fabrics made from filaments of cellulose extracted from the bacterium *Acetobacter* ; 'Microbial "tea"-shirts' (2010) Microbiology Today, 37, p. 213.

24. 'Ardil: A New British Synthetic Fibre', Draft of a report on Ardil, 12 December 1944. Astbury Papers MS419 Box E.84, ULSC BL.

25. Wood, 1950.

26. Artificial protein fibres had in fact been produced before, but with little success and never from vegetable proteins. In 1894, a fibre had been produced from gelatine, but was unsuccessful as its water resistance was poor. By the beginning of the twentieth century, attempts had been made by the chemist Todenhaupt to use the animal protein casein, which were continued later in Italy in the 1930s, but calculations showed that the amount of milk needed made the process uneconomical. Also, in both these cases, nothing was understood about the molecular changes that the process involved. The significance of Ardil was that, thanks to Astbury, Chibnall, and Bailey, it was the first time that the formation of fibrous proteins from globular ones was understood in terms of a change in molecular shape.

27. 'Ardil: a new British synthetic fibre' (1944).

28. 'X-rays and the study of textile fibres' (1942).

29. Letter from W. F. Lutyens to W. T. Astbury, 22 December 1942. Astbury Papers MS419 Box E.82, ULSC BL.

30. Letter from W. T. Astbury to W. F. Lutyens, 23 December 1942. Astbury Papers MS419 Box E.82, ULSC BL.
31. Letter from W. T. Astbury to D. Traill, 13 March 1944. Astbury Papers MS419 Box E.82, ULSC BL.
32. Letter from D. Traill to W. T. Astbury, 27 March 1944. Astbury Papers MS419 Box E.82, ULSC BL.
33. Letter from W. T. Astbury to D. Traill, 29 March 1944. Astbury Papers MS419 Box E.82, ULSC BL.
34. Letter from W. T. Astbury to D. Trail, 13 March 1944. Astbury Papers MS419 Box E.82, ULSC BL.
35. 'X-rays and the study of textile fibres' (1942).
36. Traill, 1945; p. 60.
37. Letter from W. T. Astbury to J. Weir, 8 March 1945. Astbury Papers MS419 Box E.85, ULSC BL.
38. Letter from J. Weir to J. B. Speakman, March 1945. Astbury Papers MS419 Box E. 85, ULSC BL.
39. Letter from W. T. Astbury to *Nature*, 13 December 1944. Astbury Papers MS419 Box E.82, ULSC BL.
40. Ibid.
41. Astbury, 1945a.
42. Letter from D. Traill to W. T. Astbury, 22 March 1945. Astbury Papers MS419 Box E.85, ULSC BL.
43. Letter from W. T. Astbury to D. Traill, 23 March 1945, Astbury Papers MS419 Box E.85, ULSC BL.
44. Letter from D. Traill to W. T. Astbury, 30 December 1949. Astbury Papers MS419 Box E.88, ULSC BL.
45. Letter from D. Traill to W. T. Astbury, 23 November 1950. Astbury Papers MS419 Box E.88, ULSC BL.
46. Letter from W. T. Astbury to D. Traill, 25 November 1950. Astbury Papers MS419 Box E.88, ULSC BL.
47. Letter from D. Traill to W. T. Astbury. 5 October 1951. Astbury Papers MS419 Box E.89, ULSC BL.
48. Wood, 1950.
49. Ibid.
50. 'Monkeynuts made his overcoat', Sheffield Telegraph, 1 January 1948. Astbury Papers, MS419 Box A.1 Press clippings, ULSC BL.
51. '£21,230 Gift to Leeds Professor—Aid for man who made overcoat from monkey nuts', Yorkshire Evening Post, 26 April 1948. Astbury Papers, MS419 Box A.1 Press clippings, ULSC BL.
52. Letter from W. T. Astbury to John Brocks, Manager 'Ardil' Fibre Department, 5 March 1952. Astbury Papers MS419 Box E.89, ULSC BL.

53. Ibid.
54. For an insightful and much more detailed discussion of the film and its themes, see Kemp, 1991 and Sargeant, 2008.
55. Letter from W. T. Astbury to the Vice-Chancellor, University of Leeds, 16 June 1955. Astbury Papers, MS419 Box B.18, ULSC BL.
56. Astbury, 1948; p. 271.
57. Ibid; p. 279.
58. Ibid; p. 272.
59. Ibid; p. 271.
60. Ibid; p. 275.
61. Ibid; p. 276.
62. Ibid; p. 270.
63. Judson, 1996; p. 612.
64. Ibid; p. 613.
65. Ibid; p. 614.
66. Chargaff, 1963; p. 176.
67. Chargaff, 1978; p. 3.
68. Strathern, 1998; p. 74.
69. Ibid; p. 1.
70. Ibid.
71. Chargaff, 1963; p. 199.
72. Bernal, J. D. (1949) 'New frontiers of the mind', in *The Freedom of Necessity*. London: Routledge & Kegan Paul, pp. 314–320. Cited in Brown (2005), p. 267.
73. Bernal, J. D. (1955) 'Notes on Helsinki World Assembly for Peace (22–29 June)', JDB papers, E.2.9, cited in Brown, 2005; p. 414.
74. Bernal, J. D. (1937) 'Dialectical materialism and modern science', *Science and Society*, 2, pp. 58–66. Cited in Bud, 2013; p. 316.
75. Bud, 2013; p. 312.
76. Astbury, 1948; p. 270.
77. Ibid; p. 278.
78. Astbury, 1958.
79. Chargaff, 1963; p. 198.
80. Astbury, 1948; p. 277.
81. Bud, 2013; p. 311.
82. Snow, 1964; p. 74.
83. Bud, 2013; p. 316.
84. Crick, 1966/2004; p. 93.
85. Lewis, 1945; p. 531.
86. Letter from Mrs W. Jacobs to W. T. Astbury, 14 January 1944. Astbury Papers, MS419 Box D.4, ULSC BL.

87. Letter from W. T. Astbury to Mrs W. Jacobs, 26 January 1944. Astbury Papers, MS419 Box D.4, ULSC BL.
88. Max Perutz to Gerald Holton, 9 July 1996, PRTZ 3/2/11, Archives of Churchill College, Cambridge.
89. Shang et al., 2020.
90. Walls et al., 2020.
91. Alam and Higgins, 2020.

Appendix A: Florence Bell—The 'Housewife' with X-ray Vision

1. See Williams, G. (2020) *Unravelling the double-helix*, Weidenfeld & Nicholson, p. ix. See also 'The Secret Lives of DNA' by Neeraja Sankaran, *Inference*, 6(1), June 2021 for an interesting and alternative interpretation of Franklin's comment. https://inference-review.com/article/the-secret-lives-of-dna
2. Preston, cited in Witkowski, 1980; p. 212.
3. Astbury, 1952; p. B228.
4. Williams, 2020; p. 290.
5. Cobb, 2016.
6. Ibid.
7. Watson and Crick, 1953b.
8. Judson, 1996; p. 637.
9. W. T. Astbury to Sir Charles Martin, 17 Feb 1941; MS419 E115, Papers of William Astbury, University of Leeds Brotherton Library, Special Collections Richard Synge to Robert Olby, 1st July 1980; A209, The papers and correspondence of Richard Laurence Millington Synge. Trinity College Library, Cambridge. GBR/0016/SYNG. Synge papers
10. For more on Astbury's work with Martin and Synge, and the importance of their development of partition chromatography, see Hall, 2019.
11. Synge and Martin, 1941; Mr. Chris Sawyer, personal communication.
12. Florence Ogilvy Sawyer, General Register Office, England and Wales Civil Registration Death Index, 1916–2007, London, England: Registration district: Hereford; Registration number B30D; District and Subdistrict 5161B.
13. Mr Chris Sawyer, personal communication.

Appendix B: Laszlo Lorand—The Biochemist Who Came in From the Cold

1. Professor Laszlo Lorand, personal communication.
2. Ibid.
3. Lorand, L. 2005. Factor XIII and the clotting of fibrinogen: from basic research to medicine. Journal of Thrombosis and Haemostasis, 3: 1337–1348.
4. Astbury, W.T. To Straub, F.B., 30th September 1949, Astbury Papers MS419 E108 ULSC BL.

5. W.T.Astbury to J.W. Loach, 25th Jan 1950, Astbury Papers MS419 B5 ULSC BL.

6. Lorand, L., 'Biophysical and Biochemical Studies of the Clotting of Blood,' PhD Thesis, University of Leeds, 1951.

7. Bailey, K., Bettelheim, F.R., Lorand, L., Middlebrook, W.R. 1951. 'Action of Thrombin in the Clotting of Fibrinogen.' Nature 167: 233–34.

8. Astbury, W.T., to Lorand, L., 28th January 1957, Astbury Papers MS419 E108 ULSC BL.

9. Lorand, L. to W.T. Astbury, 17th January 1957, Astbury Papers MS419 E108 ULSC BL.

10. Dr. Michele Lorand, personal communication.

11. Ibid.

Appendix C: George Washington Carver – 'the Peanut Man'?

1. Carver, 1917; p.3.

2. Ibid; pp.7-32.

3. McMurry, 1981; p.172.

4. 'Cosmetic and Process of Producing the Same' George Washington Carver of Tuskegee, Alabama. 6th Jan 1925 United States Patent Office 1522176. Online at: https://patentimages.storage.googleapis.com/8b/88/87/718c2cfc503e74/US1522176.pdf

5. See https://www.tuskegee.edu/support-tu/george-washington-carver/carver-peanut-products

6. McMurry, 1981; p.171.

7. Ibid; pp.307-313.

8. Eckelbarger et al. 2021.

Bibliography

Abir-Am, P. (1982) 'The discourse of physical power and biological knowledge in the 1930s: a reappraisal of the Rockefeller Foundation's policy in molecular biology', *Social Studies in Science*, 12, pp. 341–382.

Alam, N. and Higgins, M. K. (2020) 'A spike with which to beat covid?', *Nature Reviews: Microbiology*, 18, p. 414.

Alexander, H. E. and Leidy, G. (1951a) 'Determination of inherited traits of *H. influenzae* by desoxyribonucleic acid fractions isolated from type-specific cells', *Journal of Experimental Medicine*, 93, pp. 345–359.

Alexander, H. E. and Leidy, G. (1951b) 'Induction of heritable new type in type-specific strains of *Hemophilus influenzae*', *Proceedings of the Society for Experimental Biology and Medicine*, 78, pp. 625–626.

Alexander, H. E. and Leidy, G. (1953) 'Induction of streptomycin resistance in sensitive *Hemophilus influenzae* by extracts containing desoxyribosenucleic acid from resistant *Hemophilus influenzae*', *Journal of Experimental Medicine*, 97, pp. 17–31.

Alloway, J. L. (1932) 'The transformation in vitro of R pneumococci into S forms of different specific types by the use of filtered pneumococcus extracts', *Journal of Experimental Medicine*, 55, pp. 91–99.

Amsterdamska, O. (1993) 'From pneumonia to DNA: the research career of Oswald T. Avery', *Historical Studies in the Physical and Biological Sciences*, 24, pp. 1–40.

Anderson, L. (1988) *The story of WIRA: 70 years of wool textile research and services*. Leeds: WIRA Technology Group Ltd.

Astbury, W. T. (1923) 'The crystalline structure and properties of tartaric acid', *Proceedings of the Royal Society of London: Series A*, 102, pp. 506–528.

Astbury, W. T. (1933a) 'Some problems in the X-ray analysis of the structure of animal hairs and other protein fibres: the structure of keratins', *Transactions of the Faraday Society*, 29, p. 193.

Astbury, W. T. (1933b) 'X-ray analysis of fibres', *Nature*, 132, p. 193.

Astbury, W. T. (1934) 'X-ray studies of protein structure', *Cold Spring Harbor Symposia on Quantitative Biology*, 2, pp. 15–27.

Astbury, W. T. (1938) 'X-ray adventures among the proteins. 4th Spiers Memorial Lecture, 1937', *Transactions of the Faraday Society*, 34, pp. 378–388.

Astbury, W. T. (1939a) 'Protein and virus studies in relation to the problem of the gene', *International Conference on Genetics*, 7, pp. 49–51.

Astbury, W. T. (1939b) 'X-ray studies of the structure of compounds of biological interest', *Annual Review of Biochemistry*, 8, pp. 113–131.

Astbury, W. T. (1942) 'Sir William Bragg', *Nature*, 149, pp. 347–348.

Astbury, W. T. (1945a) 'Artificial protein fibres: their conception and preparation', *Nature*, 155, pp. 501–503.

Astbury, W. T. (1945b) 'The forms of biological molecules', in Le Gros Clark, W. E. and Medawar, P. B. (eds.) *Essays on growth and form presented to D'Arcy Wentworth Thomson*. Oxford: Oxford University Press, pp. 309–354.

Astbury, W. T. (1947) 'X-ray studies of nucleic acids', *Symposium of the Society for Experimental Biology*, 66, pp. 66–76.

Astbury, W. T. (1948) 'Science in relation to the community. The Science and Citizenship Lecture, 1 January 1948', *School Science Review*, 109, pp. 269–280.

Astbury, W. T. (1951) 'Hairs, muscles and bacterial flagella', *Nature*, 167, pp. 880–881.

Astbury, W. T. (1952) 'Adventures in molecular biology (The Harvey Lecture, 1950)', *Harvey Society Series*, 46, pp. 3–44.

Astbury, W. T. (1955) 'In praise of wool', in Proceedings of the International Wool Textiles Research Conference, *Vol. B*. Melbourne: Commonwealth Scientific and Industrial Research Organization, pp. 220–243, 220.

Astbury, W. T. (1958) 'Giant molecules', *Nature*, 181, pp. 94–95.

Astbury, W. T. (1960) 'The fundamentals of fibre research: a physicist's story', *Journal of the Textile Industry*, 51, pp. 515–525.

Astbury, W. T., Bailey, K., and Rudall, K. M. (1943) 'Fibrinogen and fibrin as members of the keratin-myosin group', *Nature*, 151, pp. 716–717.

Astbury, W. T. and Bell, F. O. (1938a) 'Some recent developments in the X-ray study of proteins and related structures', *Cold Spring Harbor Symposia on Quantitative Biology*, 6, pp. 109–118.

Astbury, W. T. and Bell, F. O. (1938b) 'X-ray studies of thymonucleic acid', *Nature*, 141, pp. 747–748.

Astbury, W. T. and Bell, F. O. (1941) 'The nature of the intramolecular fold in alpha-keratin and alpha-myosin', *Nature*, 147, pp. 696–699.

Astbury, W. T. and Dickinson, S. (1935a) 'α–β intramolecular transformation of muscle protein in situ', *Nature*, 135, p. 765.

Astbury, W. T. and Dickinson, S. (1935b) 'α–β intramolecular transformation of myosin', *Nature*, 135, p. 95.

Astbury, W. T. and Dickinson, S. (1936) 'X-ray study of myosin', *Nature*, 137, pp. 909–910.

Astbury, W. T. and Dickinson, S. (1940) 'X-ray studies of the molecular structure of myosin', *Proceedings of the Royal Society of London, Series B*, 129, pp. 307–332.

Astbury, W.T., Dickinson, S., and Bailey, K. (1935) 'X-ray interpretation of denaturation and the structure of seed globulins', *The Biochemical Journal*, 29, p. 2351.

Astbury, W. T., and Lomax, R. (1934) 'X-ray photographs of crystalline pepsin', *Nature*, 133, p. 795.

Astbury, W. T., and Lomax, R. (1935) 'An X-ray study of the hydration and denaturation of proteins', *Journal of the Chemical Society*, pp. 846–51.

Astbury, W. T. and Preston, R. D. (1940) 'The structure of the cell wall in some species of the filamentous green alga cladophora', *Proceedings of the Royal Society of London, Series B*, 129, pp. 54–76.

Astbury, W. T. and Sisson, W. A. (1935) 'X-ray studies of the structure of hair, wool and related fibres. III. The configuration of the keratin molecule and its orientation in the biological cell', *Proceedings of the Royal Society of London, Series A*, 150, pp. 533–551.

Astbury, W. T. and Street, A. (1932) 'The X-ray studies of the structure of hair, wool, and related fibres. I. General', *Philosophical Transactions of the Royal Society of London, Series A*, 230, pp. 75–101.

Astbury, W. T. and Weibull, C. (1949) 'X-ray diffraction study of the structure of bacterial flagella', *Nature*, 163, pp. 280–281.

Astbury, W. T. and Woods, H. J. (1931a) 'Report of the Work Done under the Research Scheme Established in 1928 with the Aid of a Special Grant from the Worshipful Company of Clothworkers', Session 1930–1931, pp. 16–25.

Astbury, W. T. and Woods, H. J. (1931b) 'The molecular weight of proteins', *Nature*, 127, pp. 663–665.

Astbury, W. T. and Woods, H. J. (1932) 'The molecular structure of textile fibres', *Journal of the Textile Industry*, 23, p. T17.

Astbury, W. T. and Woods, H. J. (1933) 'X-ray studies of the structure of hair, wool and related fibres. II. The molecular structure and elastic properties of hair keratin', *Philosophical Transactions of the Royal Society A*, 232, pp. 333–394.

Avery, O. T., Macleod, C M., and McCarty, M. (1944) 'Studies on the chemical nature of the substance inducing transformation of pneumococcal types: induction of transformation by a desoxyribonucleic acid fraction isolated from pneumococcus type III', *Journal of Experimental Medicine*, 79, pp. 137–158.

Bailey, K. (1961) 'William Thomas Astbury 1898–1961: a personal tribute', *Advances in Protein Chemistry*, 17, pp. x–xiv.

Bailey, K., Bettelheim, F. R., Lorand, L., and Middlebrook, W. R. (1951) 'Action of thrombin in the clotting of fibrinogen', *Nature*, 167, pp. 233–234.

Bawden, F. C., Pirie, N. W., Bernal, J. D., and Fankuchen, I. (1936) 'Liquid crystalline substances from virus-infected plants', *Nature*, 138, pp. 1051–1052.

Bennett, A. (1902) *Anna of the five towns*. London: Penguin Books.

Bennett, A. (1908) *The old wives' tale*. London: Penguin Books.

Berg, P., Baltimore, D., Boyer, H., Cohen, S., Davis, R., Hogness, D., Nathans, D., Roblin, R., Watson, J. D., Weissman, S., and Zinder, N. D., (1974) 'Potential biohazards of recombinant DNA molecules', *Science*, 185, p. 303.

Bernal, J. D. (1963) 'William Thomas Asbury 1898–1961', *Biographical Memoirs of Fellows of the Royal Society*, 9, pp. 1–35.

Bernal, J. D. and Crowfoot, D. M. (1934) 'X-ray photographs of crystalline pepsin', *Nature*, 133, pp. 794–795.

Boivin, A. (1947) 'Directed mutation in colon bacilli, by an inducing principle of desoxyribonucleic nature: its meaning for the general biochemistry of heredity', *Cold Spring Harbor Symposia on Quantitative Biology*, 12, pp. 7–17.

Bradford, E. (2008) *Headingley: 'this pleasant rural village'*. Huddersfield: Northern Heritage Publications/ Jeremy Mills Publishing.

Bragg, W. L. (1912) 'The diffraction of short electromagnetic waves by a crystal', *Proceedings of the Cambridge Philosophical Society*, 17, pp. 43–57.

Bragg, W. H. (1920) *The world of sound*. London: G. Bell and Sons.

Bragg, W. L. (1926) 'The imperfect crystallisation of common things', *Nature*, 118, pp. 120–122.

Bragg, W. H. and Bragg, W. L. (1913) 'The reflection of X-rays by crystals', *Proceedings of the Royal Society of London, Series A*, 88, pp. 428–438.

Bragg, W. L. and Caroe, G. G. (1962) 'Sir William Bragg, F.R.S. (1862–1942)', *Notes and Records of the Royal Society*, 17, pp. 162–182.

Brown, A. (2005) *J. D. Bernal: the sage of science*. Oxford: Oxford University Press.

Bud, R. (2013) 'Life, DNA and the model', *The British Journal for the History of Science*, 46, pp. 311–334.

Burt, S. and Grady, K. (1994) *The illustrated history of Leeds*. Derby: Breedon Books.

Caroe, G. (1978) *William Henry Bragg 1862–1942, man and scientist*. Cambridge: Cambridge University Press.

Carver, G.W. (1917) 'How to Grow the Peanut and 105 Ways of Preparing It For Human Consumption' *Tuskegee Institute Research Station Bulletin* 31.

Chargaff, E. (1950) 'Chemical specificity of nucleic acids and mechanism of their enzymatic degradation', *Experientia*, 6, pp. 201–240.

Chargaff, E. (1963) *Essays on nucleic acids*. Amsterdam: Elsevier Publishing Company.

Chargaff, E. (1978) *Heraclitean fire: sketches from a life before nature*. New York: The Rockefeller University Press.

Chargaff, E. and Vischer, E. (1948a) 'The composition of the pentose nucleic acids of yeast and pancreas', *Journal of Biological Chemistry*, 176, pp. 715–734.

Chargaff, E. and Vischer, E. (1948b) 'The separation and quantitative estimation of purines and pyrimidines in minute amounts', *Journal of Biological Chemistry*, 17, pp. 703–714.

Chargaff, E., Vischer, E., Doniger, R., Green, C., and Misani, F. (1949) 'The composition of the desoxypentose nucleic acids of thymus and spleen', *Journal of Biological Chemistry*, 177, pp. 405–416.

Chargaff, E. and Zamenhof, S. (1948) 'The isolation of highly polymerized desoxypentosenucleic acid from yeast cells', *Journal of Biological Chemistry*, 173, pp. 327–335.

Chargaff, E., Zamenhof, S., and Green, C. (1950) 'Composition of human desoxypentose nucleic acid', *Nature*, 165, pp. 756–757.

Chibnall, A. C. (1964) 'Kenneth Bailey, 1909–1963', *Biographical Memoirs of Fellows of the Royal Society*, 10, pp. 1–13.

Clarke, I. (2008) *Molecular biology and publics: the case of Astbury at Leeds*. MA thesis, University of Leeds.

Cobb, M. (2016) 'A speculative history of DNA: what if Oswald Avery had died in 1934?', *PLOS Biology*, 14, e2001197.

Cobourn, A. F. (1969) 'Oswald Theodore Avery and DNA', *Perspectives in Biology and Medicine*, 12, pp. 623–630.

Cochran, W., Crick, F. H. C., and Vand, V. (1952) 'The structure of synthetic polypeptides, I. The transform of atoms on a helix', *Acta Crystallographica*, 5, pp. 581–586.

Cohen, J. S. and Portugal, F. H. (1977) *A century of DNA: a history of the discovery of the structure and function of the genetic substance*. Cambridge: MIT Press.

Commoner, B. (1968) 'Failure of the Watson–Crick theory as a chemical explanation of inheritance', *Nature*, 220, pp. 334–340.

Consden, R., Gordon, A. H., and Martin, A. J. P. (1944) 'Qualitative analysis of proteins: a partition chromatographic method using paper', *The Biochemical Journal*, 38, pp. 224–232.

Creeth, J. M., Gulland, J. M., and Jordan, D. O. (1947) 'Deoxypentose nucleic acids. Part III. Viscosity and streaming birefringence of solutions of the sodium salt of the deoxypentose nucleic acid of calf thymus', *Journal of the Chemical Society*, 0, 1141–1145.

Crick, F. H. C. (1958) 'On protein synthesis', *Symposia of the Society for Experimental Biology*, 12, pp. 138–163.

Crick, F. H. C. (1966/2004) *Of molecules and men*. Amherst: Prometheus Books.

Crick, F. H. C. (1988) *What mad pursuit: a personal view of scientific discovery*. London: Weidenfeld and Nicholson.

Cushing, D. (2005) 'Reginald Dawson Preston', *Biographical Memoirs of Fellows of the Royal Society*, 51, pp. 348–353.

Davies, M. (1990) 'W. T. Astbury, Rosie Franklin, and DNA: a memoir', *Annals of Science*, 47, pp. 607–618.

Dawson, M. H. and Sia, R. H. (1931a) 'In vitro transformation of pneumococcal types: I. A technique for inducing transformation of pneumococcal types in vitro', *Journal of Experimental Medicine*, 54, pp. 681–699.

Dawson, M. H. and Sia, R. H. (1931b) 'In vitro transformation of pneumococcal types: II. The nature of the factor responsible for the transformation of pneumococcal types', *Journal of Experimental Medicine*, 54, pp. 701–710.

Dickson, D. (1980) 'Genentech makes splash on Wall Street', *Nature*, 287, p. 669.

Dowling, H. F. (1977) *Fighting infection: conquests of the twentieth century*. Cambridge: Harvard University Press.

Dubos, R. J. (1976) *The professor, the institute, and DNA*. New York: The Rockefeller University Press.

Eckelbarger, M., Rice, S.L., Osano, A., Peng, J., Ullah, H., Rhee, S-Y. (2021) 'Recognising pioneering Black plant scientists in our schools and society' *Trends in Plant Science*, 26, pp. 989–992.

Fara, P. (2009) *Science: a four-thousand-year history*. Oxford: Oxford University Press.

Ferry, G. (2008) *Max Perutz and the secret of life*. London: Pimlico.

Forster, G. C. F. (1980) 'The foundations: from the earliest times to c. 1700', in Fraser, D. (ed.) *A history of modern Leeds*. Manchester: Manchester University Press, 5.

Fox Keller, E. (2000) *The century of the gene*. Cambridge: Harvard University Press.

'German industry and the war', (1918) *Nature*, 102, pp. 66–67.

Gilleghan, J. (2001) *Leeds: a to z of local history*. Leeds: Kingsway Press.

Glynn, J. (2012) *My sister Rosalind Franklin*. Oxford: Oxford University Press.

Goertzel, T. and Goertzel, B. (1995) *Linus Pauling: a life in science and politics*. New York: Basic Books.

Goldschmidt, L., Teng, P. K., Riek, R., and Eisenberg, D. (2010) 'Identifying the amylome, proteins capable of forming amyloid-like fibrils', *Proceedings of the National Academy of Sciences of the United States of America*, 107, pp. 3487–3492.

Gordon, Hugh. (1996) 'Richard Laurence Millington Synge, 28 October 1914–18 August 1994', *Biographical Memoirs of Fellows of the Royal Society*, 42, pp. 455–476.

Greenstein, J. P. (1943) 'Friedrich Miescher, 1844–1895', *Scientific Monthly*, 57, pp. 523–530.

Griffith, F. (1928) 'The significance of pneumococcal types', *Journal of Hygiene*, 27, pp. 135–159.

Gulland, J. M. (1947) 'The structure of nucleic acids', *Cold Spring Harbor Symposia on Quantitative Biology*, 12, pp. 95–103.

Gulland, J. M., Barker, G. R., and Jordan, D. O. (1945) 'The chemistry of the nucleic acids and nucleoproteins', *Annual Review Biochemistry*, 14, pp. 175–206.

Gulland, J. M., Jordan, D. O., and Taylor, H. F. (1947) 'Deoxypentose nucleic acids. Part II. Electrometric titration of the acidic and the basic groups of the deoxypentose nucleic acid of calf thymus', *Journal of the Chemical Society*, 0, 1131–1141.

Hager, T. (1995) *Force of nature: the life of Linus Pauling*. New York: Simon and Schuster.

Hall, K.T. (2019) 'In praise of wool: the development of partition chromatography and its under-appreciated impact on molecular biology', *Endeavour*, 43, p. 100708.

Heaton, H. (1965) *The Yorkshire woollen and worsted industries*. 2nd edn. Oxford: Clarendon Press.

Hershey, A. D. and Chase, M. (1952) 'Independent functions of viral protein and nucleic acid in growth of bacteriophage', *Journal of General Physiology*, 36, pp. 39–56.

Hodgkin, D. M. C. (1975) 'Kathleen Lonsdale: 28 January 1903–1 April 1971', *Biographical Memoirs of Fellows of the Royal Society*, 21, pp. 447–484.

Hodgkin, D. M. C. (1980) 'John Desmond Bernal: 10 May 1901–15 September 1971', *Biographical Memoirs of Fellows of the Royal Society*, 26, pp. 16–84.

Hotchkiss, R. (1955) 'Bacterial transformation', *Symposium on Genetic Recombination. Journal of Cellular and Comparative Physiology*, 45(Suppl 2), pp. 1–21.

Hunter, G. K. (2000) *Vital forces: the discovery of the molecular basis of life*. London: Academic Press.

Jackson, D. A., Symonds, R. H., and Berg, P (1972) 'Biochemical method for inserting new genetic information into DNA of simian virus 40: circular SV40 DNA molecules containing lambda phage genes and the galactose operon of *Escherichia coli*', *Proceedings of the National Academy of Sciences of the United States of America*, 69, pp. 2904–2909.

Jamieson, A. (2013) 'More than meets the eye: revealing the therapeutic potential of "light"', *Social History of Medicine*, 26, pp. 715–737.

Jenkin, J. 2011. *William and Lawrence Bragg, father and son—the most extraordinary collaboration in science.* Oxford: Oxford University Press.

Jones, R. and Lodge, O. (1896) 'The discovery of a bullet lost in the wrist by means of the Roentgen rays', *The Lancet*, 147, pp. 476–477.

Judson, H.F. 1980. 'Reflections on the historiography of molecular biology', *Minerva*, 18, pp. 369–421.

Judson, H. F. (1996) *The eighth day of creation*. Cold Spring Harbor: Cold Spring Harbor Press.

Kellgren, J. H., Ball, J., Reed, R., Beighton, E., and Astbury, W. T. (1951) 'Biophysical studies of rheumatoid connective tissue', *Nature*, 168, pp. 493–494.

Kelly, T. J. and Smith, H. O. (1970) 'A restriction enzyme from *Hemophilus influenzae*: II. Base sequence of the recognition site', *Journal of Molecular Biology*, 51, pp. 393–400.

Kemp, P. (1991) *Lethal innocence: the cinema of Alexander Mackendrick*. London: Methuen.

Klug, A. (1974) 'Rosalind Franklin and the double helix', *Nature*, 248, pp. 787–788.

Konforti, B. (2000) 'The servant with the scissors', *Nature Structural Biology*, 7, pp. 99–100.

Lederberg, J. (1972) 'Reply to Wyatt', *Nature*, 239, p. 234.

Lederberg, J. (2000) 'The dawning of molecular genetics', *Trends in Microbiology*, 8, pp. 194–95.

Leidy, G., Hahn, E., and Alexander, H. E. (1953) 'In vitro production of new types of *Hemophilus influenzae*', *Journal of Experimental Medicine*, 97, pp. 467–482.

Lewis, C. S. (1945) *That hideous strength*. Oxford: Bodley Head.

Lonsdale, K. (1961) 'Prof. W. T. Astbury, F.R.S.', *Nature*, 191, p. 332.

Lorand, L. (2005) 'Factor XIII and the clotting of fibrinogen: from basic research to medicine', *Journal of Thrombosis and Haemostasis*, 3, pp. 1337–1348.

Lovelock, J. (2004) 'Archer John Porter Martin CBE: 1 March 1910–28 July 2002', *Biographical Memoirs of Fellows of the Royal Society*, 50, pp. 157–170.

Maddox, B. (2002) *Rosalind Franklin: The dark lady of DNA*. London: Harper Collins.

Manchester, K. (1995) 'Did a tragic accident delay the discovery of the double-helical structure of DNA?', *Trends in Biochemical Sciences*, 20, pp. 126–128.

Martin, AJP, and RLM Synge. (1941) 'A new form of chromatogram employing two liquid phases. 1. Theory of chromatography. 2. Application to the micro determination of the higher mono amino acids in proteins', *The Biochemical Journal*, 35, pp. 1358–1368.

McMurry, L.O. (1981) *George Washington Carver: Scientist and Symbol*. Oxford University Press.

Meldola, R. (1915) 'Professional chemists and the war', *Nature*, 95, pp. 18–19.

Meyer, H. K. and Mark, H. (1928) 'Uber den Bau des Krystallisierten Anteils der Cellulose', *Berichte den Deutschen Chemisches Gesellschaft*, 61, pp. 593–614.

Midgley, M. (2001) *Science and poetry*. London: Routledge.

Moore, G. M. (1989) *Schrödinger: life and thought*. Cambridge: Cambridge University Press.

Morange, M. (1998) *A history of molecular biology*. Cambridge: Harvard University Press.

Morgan, W. T. J. (1944) 'Transformation of pneumococcal types', *Nature*, 153, pp. 763–764.

Nelson, B. (1980) *The woollen industry of Leeds*. Leeds: D. & J. Thornton.

Newmark, P. (1983) 'Thirty years of DNA', *Nature*, 305, p. 383.

Olby, R. (1972) 'Avery in retrospect', *Nature*, 239, pp. 295–296.

Olby, R. (1979) 'Mendel - no Mendelian?', *History of Science*, 17, pp. 53–72.

Olby, R. (1984) 'The sheriff and the cowboys: or Weaver's support of Astbury and Pauling', *Social Studies in Science*, 14, pp. 244–247.

Olby, R. (1994) *The path to the double helix: the discovery of DNA*. New York: Dover Publications.

Olby, R. (2003) 'Quiet debut for the double helix', *Nature*, 421, pp. 402–405.

Olby, R. (2009) *Francis Crick: hunter of life's secrets*. Cold Spring Harbor: Cold Spring Harbor Press.

Passey, R. D., Dmochowski, L., Reed, R., and Astbury, W. T. (1950) 'Biophysical studies of extracts of tissues from high- and low-breast-cancer strain mice', *Biochemica et Biophysica Acta*, 4, pp. 391–409.

Passey, R. D., Dmochowski, L., Reed, R., Astbury, W. T., and Johnson. P. (1950) 'Electron microscope studies of normal and malignant tissues of high- and low-breast-cancer strain mice', *Nature*, 165, p. 107.

Passey, R. D., Dmochowski, L., Reed, R., Eaves, G., and Astbury, W. T. (1951) 'Electron microscope studies of human breast cancer', *Nature*, 167, pp. 643–644.

Pauling, L. and Corey, R. (1953) 'A proposed structure for the nucleic acids', *Proceedings of the National Academy of Sciences of the United States of America*, 39, pp. 84–97.

Pauling, L., Itano, H. A., Singer, S. J., and Wells, I. C. (1949). 'Sickle cell anaemia, a molecular disease', *Science*, 110, pp. 543–548.

Phillips, D. 1979. 'William Lawrence Bragg. 31 March 1890–1 July 1971', *Biographical Memoirs of Fellows of the Royal Society*, 25, pp. 74–143.

'The physicist in the textile industries', (1923) *Nature*, 112, pp. 707–708.

Pohl, R. (1909) 'The defects of English technical education and the remedy', *Nature* 80: 205–08.

Preston, R. D. (1974) 'William Thomas Astbury, F.R.S. - fibrous polymer extraordinary', in Atkins, E. D. T. and Keller, A. (eds.) *Structure of fibrous biopolymers: proceedings of the twenty-sixth symposium of the Colston Research Society* London: Butterworths, p. 1.

'The promotion of textile industries', (1918) *Nature*, 102, p. 98.

Redfearn, J. 1980. 'Biogenetic hormones: insulin trial', *Nature*, 286, p. 436.

Ridley, M. (2006) *Francis Crick: discoverer of the genetic code*. New York: Harper Collins.

Röntgen, W. C. (1896) 'On a new kind of rays', *Nature*, 53, pp. 274–276.

Rose, S. 1997. *Lifelines: Biology, Freedom, Determinism*. London: Penguin Books.

Rossi, H. H. and Kellerer, A. M. (1995) 'Röntgen', *Radiation Research*, 144, pp. 124–128.

Sanger, F. (1952) 'The Arrangement of Amino Acids in Proteins', *Advances in Protein Chemistry*, 7, pp. 1–67.

Sanger, F. (1988) Sequences, Sequences, Sequences, *Annual Review of Biochemistry*, 57, pp. 1–29.

Sargeant, A. 2008. '*The Man in the White Suit*: new textiles and the social fabric', *Visual Culture in Britain*, 9, pp. 27–53.

Sayre, A. (1975) *Rosalind Franklin and DNA*. New York: Norton.

Schultz, J. (1941) 'The evidence for the nucleoprotein nature of the gene', *Cold Spring Harbor Symposia in Quantitative Biology*, 9, pp. 55–65.

'Scientific and Industrial Research', (1917) *Nature*, 100, pp. 17–20.

Serafini, A. 1989. *Linus Pauling: A Man and His Science*. New York: Simon and Schuster.

Shang, J., Ye, G., Shi, K., Wan, Y., Luo, C., Aihara, H., Geng, Q., Auerbach, A., and Li, F (2020) 'Structural basis of receptor recognition by SARS-CoV-2', *Nature*, 581, pp. 221–224.

Signer, R., Caspersson, T., and Hammarsten. E. (1938) 'Molecular shape and size of thymonucleic acid', *Nature*, 141, p. 122.

Simmons, C. 1910. 'Technological science in Germany', *Nature*, 82, pp. 313–314.

Smith, H. O. and Wilcox, K. W. (1970) 'A restriction enzyme from *Hemophilus influenzae*: I. purification and general properties', *Journal of Molecular Biology*, 51(2), pp. 379–391. https://doi.org/10.1016/0022-2836(70)90149-X

Snow, C. P. (1964) *The two cultures*. 2nd edn. Cambridge: Cambridge University Press.

Stanley, W.M. 1935. 'Isolation of a crystalline tobacco-mosaic virus protein', *Science* 81: 644–45.

'The state and industrial research', (1927) *Nature*, 119, pp. 589–591.

Stent, G. (1972) 'Prematurity and uniqueness in scientific discovery', *Scientific American*, 227, pp. 84–93.

Stent, G. S. (1966) 'Waiting for the paradox', in Cairns, J., Stent, G. S., and Watson, J. D. (eds.) *Phage and the origins of molecular biology*, Cold Spring Harbor: Cold Spring Harbor Laboratory Press.

Stern, K. G. (1947) 'Nucleoproteins and gene structure', *Yale Journal of Biological Medicine*, 19, pp. 937–49.

Strathern, P. (1998) *Oppenheimer and the bomb*. London: Arrow Books.

Synge, R. L. M. and Martin, A. J. P. (1941) 'Some applications of periodic acid to the study of the hydroxyamino-acids of protein hydrolysates: the liberation of acetaldehyde and higher aldehydes by periodic acid. 2. Detection and isolation of formaldehyde liberated by periodic acid. 3. Ammonia spli', *The Biochemical Journal*, 35, pp. 294–314.

Thornton, D. (2002) *Leeds: the story of a city*. Ayr: Fort Publishing.

Traill, D. (1945) 'Vegetable proteins and synthetic fibres', *Chemistry and Industry*, February 24, 58–63.

Uversky, V. N. and Fink, A. L. (eds.) (2007) *Protein misfolding, aggregation, and conformational disease, part B: molecular mechanisms of conformational diseases*. New York: Springer.

Vischer, E. and Chargaff, E. (1947) 'The separation and characterization of purines in minute amounts of nucleic acid hydrolysates', *Journal of Biological Chemistry*, 168, pp. 781–782.

Vischer, E., Zamenhof, S., and Chargaff, E. (1949) 'Microbial nucleic acids: the desoxypentose nucleic acids of avian tubercle bacilli and yeast', *Journal of Biological Chemistry*, 177, pp. 429–438.

Waddington, C. H. (1961a) 'Letter', *Nature*, 190, p. 184.

Waddington, C. H. (1961b) 'Letter', *Nature*, 190, p. 1124.

Walls, A. C., Park, Y-J., Tortorici, M. A., Wall, A., McGuire, A. T., and Veesler, D (2020) 'Structure, function and antigenicity of the SARS-CoV-2 spike glycoprotein', *Cell*, 180, pp. 281–192.

'The war and British chemical industry', (1915) *Nature*, 95, pp. 119–120.

Watson, J. D. (1968) *The double helix*. London: Weidenfeld & Nicholson.

Watson, J. D. and Crick, F. H. C. (1953a) 'Genetical implications of the structure of deoxyribonucleic acid', *Nature*, 171, pp. 964–967.

Watson, J. D. and Crick, F. H. C. (1953b) 'Molecular structure of nucleic acids: a structure for deoxyribose nucleic acid', *Nature* 171: 737–38.

Watson, J. D. and Tooze, J. (1981) *The DNA story: a documentary history of gene cloning*. San Francisco: W. H. Freeman and Company.

Watson, J. D., Hopkins, N. H., Roberts, J. W., Steitz, J. A., and Weiner, A. M. (1987) *Molecular biology of the gene*. 4th edn. Menlo Park: Benjamin/Cummings.

Wilkins, M. (2003) *The third man of the double-helix*. Oxford: Oxford University Press.

Wilkins, M. H. F. (1987) 'John Turton Randall. 23 March 1905–16 June 1984', *Biographical Memoirs of Fellows of the Royal Society*, 33, pp. 492–535.

Witkowski, J. A. (1980). 'W. T. Astbury and R. G. Harrison: the search for the molecular determination of form in the developing embryo', *Notes and Records of the Royal Society of London*, 35, pp. 195–219.

Wood, A. (1950) *The groundnut affair*. London: Bodley Head.

Wyatt, H. V. (1972) 'When does information become knowledge?', *Nature* 235, pp. 86–89.